Praise for *Making Futures Work*

If you want to drive futures conversations and need practical advice on where to start, this is your toolkit. Phil helps you understand how we got here, why we use these tools, and where to go beyond the edges.

—Christy Ennis Kloote, Experience Director and Advisor, Latticework

What I have seen Phil and the Futures team at McKinsey do is to truly open the eyes of an organization, at the highest levels, to think about change from a macroeconomic lens in order to create new companies, products, and services that extend their value beyond the present. We have seen too many organizations stuck in design as a cost savings mode versus the true growth, innovation, and future-proofing ability it needs. With this book, organizations can thrive and grow in scenarios grounded in real thinking that will help them plan and succeed, with substantial impact.

—Jennifer Kilian, Partner, McKinsey & Company

Balagtas offers us a mindset, framework, and methodology for futures work. Organizations of all kinds are tempted to shy away from the process of muscular, long-term thinking, often because they are more comfortable with designing iterations on past and present successes—more "SKUs" without meaningful innovation. This book gives futurists and designers an excellent set of tools to work through organizational myopia and timidity in order to help companies strategize about the future with confidence and enthusiasm.

—Brenda Laurel, Principal, Neogaian Interactive

Everyone wants to know what is going to happen in the future. Futures thinkers use multiple scenarios to answer that question. But an even more important questions is, "How can I make the future better?" Design is one of the most popular and effective ways of answering that question, and Phil Balagtas is just the person to show how anticipating what will happen is an important step in designing what should happen.

—*Peter Bishop, Founder, Teach the Future*

With design, we can stop acting in reaction to predictive events and be a proponent of creating the realities we want for our organizations or society. *Making Futures Work* is a testament and guide to how we can use futures and design more proactively and effectively and how to adapt today's thinking into more visionary leadership.

—*Juliana Proserpio, Founder and Chief Design Officer,*
Echos Desirable Futures Lab

There are many books and online resources about UX but none of them provide concrete examples and clear methodologies to cultivate innovation in an organization. *Making Futures Work* is an eye-opener on how to think outside the box to shape the future.

—*Kévin Meunier, CEO, Offinite*

Making Futures Work

Integrating Futures Thinking
for Design, Innovation, and Strategy

Phil Balagtas
Foreword by Cat Drew
Afterword by Ben Lowdon

Beijing · Boston · Farnham · Sebastopol · Tokyo

Acquisitions Editor: Amanda Quinn

Development Editor: Angela Rufino

Production Editor: Gregory Hyman

Copyeditor: Arthur Johnson

Proofreader: Sonia Saruba

Indexer: Potomac Indexing, LLC

Interior Designer: Monica Kamsvaag

Cover Designer: Susan Thompson

Illustrator: Kate Dullea

June 2024:   First Edition

Revision History for the First Edition

2024-06-05:   First Release

See *http://oreilly.com/catalog/errata.csp?isbn=9781098148904* for release details.

978-1-098-14890-4

[LSI]

Contents

Foreword

Faced with the greatest challenge yet for humanity—the twin climate and nature crises—we simply have no choice but to design for a future that will be radically different from the way we live now. We can no longer continue our extractive ways of living and consuming and seeing ourselves as separate from or "above" nature. We must live more regeneratively and equitably, creating businesses and organizations that put people and the planet at the heart of what they do, with profit in their service rather than at their detriment.

Futures Thinking is critical to this. It is not about making a faster, better, smoother version of what we already have but about fundamentally redesigning how we live, travel, eat, work, and play. And to do that, we need to be able to imagine alternatives—radical alternatives. Alternatives that have stretched our imagination to their logical conclusions, alternatives that have existed in our past but have been forgotten, alternatives that exist in our present but are undervalued. And then we need to design them into visual and physical things to bring them to life so that people can see they are possible, so we can build or grow them together.

In case you need a more hardheaded reason, futures keeps you in the game. The roots of strategic foresight go back to the military and to business strategy, of course, and the discipline is championed within the scientific professions in government. Why would you not want to understand what might be ahead of you, what emerging patterns and trends might affect your business, so you can respond? But Design allows you to move beyond strategic foresight into action and to create opportunities within (and at the edges of) an unfolding future. It puts you at the forefront of helping to shape a more sustainable and regenerative world for our grandchildren.

What I love about futures design is that it is not about predicting the future. As the pioneers of critical design, Anthony Dunne and Fiona Raby, stated in their book *Speculative Everything* (MIT Press), "In our view, [trying to pin the future down] is a pointless activity. What we are interested in, though, is the idea of possible futures and using them as tools to better understand the present and to discuss the kind of future people want, and, of course, ones people don't want"—and as Phil has reiterated, "We don't tell you the future; we offer possibilities." That is why futures design is not only a great tool for dealing with uncertainty (how can you predict anything in this ever-changing world?) but is also inherently collaborative. It helps you co-design the future with others: colleagues, stakeholders, shareholders, customers (or, better put, citizens). It invites questions and opens up the imagination space so that others can design with you, eliciting their creativity, resources, and buy-in.

In my experience of trying to embed futures, specifically Speculative Design, into government, it has often been difficult to translate the methods into workshop tools that could be used to co-design with others, which, as I just said, is one of the main benefits of futures design. Designers have created tarot cards, narrative prompts, rapid ideation methods, and future headline activities to help, but these must be context specific and often need carefully designed prompts to stretch people's understanding of what might be possible. And skilled designers are needed to translate ideas into visual prototypes that can provoke. *Making Futures Work* is therefore a helpful and necessary book for setting up the process, mindsets, and methods for futures work that designers and nondesigners can use to embed Futures Thinking within organizations, and for reiterating the importance of multidisciplinary teams, including those who have not only the subject matter expertise but also the visual design skills to make this work.

As with all types of design, what is as important as the tools and process is the mindset. Educated/worldly positivity, comfort with uncertainty, inclusive imagination—these are attributes that we all will need as we get ready to enter the second quarter of the 21st century, so that we may adapt to our changing world and design our way toward a better one.

—Cat Drew
Chief Design Officer
The Design Council

Preface

Making Futures Work is a culmination of observations, conversations, experiments, stories, and perspectives I've gathered as I've investigated Futures Thinking and tried to integrate it into my own design practice. Since I first discovered futures through a practice called Speculative and Critical Design (which I'll describe later in the book), I've been passionately curious about the process and methods to determine how, when, and where futures could be applied to serve as an approach for innovation and strategy. Further investigation into the various ways to think about the future led me to Strategic Foresight, which has been around for many years. While there are many tools out there that are futures specific or futures adjacent, I've tried to collect the fundamental principles and methods that I believe to be most useful in undertaking futures work.

What I've found over the years is that futures can be difficult to sell, advocate for, and implement. Audiences don't always understand its value, how it works, and why it matters. I've also found that the techniques can seem intimidating or complex to newcomers. Unlike traditional Design Thinking, futures is not a linear process. It is fundamentally based on the idea of pluralities—that there is no one future we are designing for, but rather a variety of alternatives that we can (and should) consider to help us understand what could happen in the short- and long-term future. However, what I've also discovered is that to make futures work, we have to know what to do with that information. Applying this process to strategic thinking and determining how to create an agenda *today* to plan toward preferable futures and divert from nonpreferable futures is the most important aspect of futuring. Without a strategy to deliver on that preferable future, you are merely having a conversation, and for some that may not be enough to incite change or deliver a measurable outcome.

Since around 2009, I've been using the various methodologies within Futures Thinking in different capacities for product design, digital and cultural transformations, education for children, training for corporate design teams, and various other projects; this includes using it every day to plan my own career and life goals. And while there's a vast amount of information and a range of approaches on this topic, I've come to several conclusions about how to practically explain, facilitate, and defend it to make it useful for many, including in my own practice as a designer and strategist. Futures Thinking isn't that different from the mental models we use every day. We are all inherently *futuring* every second. From thinking about our workday to planning product roadmaps, systematic, mechanical, analytical, and anticipatory modeling is part of our DNA and goes back to our imperative to survive in a complex and uncertain world. Futures thereby is an alternative conceptual framework and scaffolding for how we can think about potential outcomes and implications.

A question I've struggled with during my research is, why isn't futures practiced everywhere? Shouldn't everyone be thinking about long-term implications and analyzing threats and scenarios? And yet it has felt like the practice has mostly been relegated to academia, government, or corporate strategy, or has been facilitated by the few who have managed to make a career of consulting or teaching. That is slowly changing, though, and within the last decade more and more people have been using some form of Futures Thinking in a variety of organizations, situations, and contexts. This book will help demystify that process and will offer a strategy for how you might use the different tools and how you can discuss futures in a way that is flexible and accessible. Overall, *Making Futures Work* is an ode to the foundational work many futurists and designers have accomplished over the years, as well as an alternative approach to how we might use the same tools in the design community today. It's intentionally framed to consider different views of what futures is (or isn't) and laid out so that you can form your own opinion about how to make it work for you.

Who This Book Is For

This book is primarily for designers and design leaders—those with a background in digital and product design, strategy design, service design, or experience design, and who have at least a few years' experience as a practitioner. I say this because designers might not necessarily be interested in strategic thinking early in their careers. Those who have had some time to hone their craft and are looking to expand their role or capabilities with strategy and innovation processes will likely be more prepared to pick this book up and run with it. Designers are the community I have built my career around and whom I am usually speaking

to or teaching. That said, this book is really for anyone who is interested in getting started in learning about Futures Thinking. Managers, executive leaders, and other strategic thinkers can certainly benefit from the content. The book is also for those who have worked internally (in-house) and/or externally (consultants) and will provide different perspectives on how things might change if you are using futures within a team or as an outside practitioner advising an organization. Since the workforce landscape is constantly evolving, new roles are being created every day; thus I'm not limiting the audience for this book to any one type of specialization. That said, it is written from a designer's perspective and includes vocabulary and methods that designers might be more familiar with. However, I discuss many concepts that don't require a design degree to understand.

Thus, if you find value in the content, by all means use this book, whether as a quick-start guide, a workshop manual, a textbook, or a strategic process guide. It was written to be as practical and usable as possible, without extraneous commentary, philosophy, or political stances. Other than the occasional quip about the usefulness of a method or principle or the observation that futures should be used to do more good in the world, I have tried to refrain from overly pontificating about futures as a cure-all. Essentially, the book is meant to be used as a supplement to your current tools and processes. Whether you want to think more strategically or just need a few more ways to consider alternative ideas or perspectives, I hope you'll find something in here that is useful. I do not intend to claim (other than in the necessary legal copyrights of this book) any stake in or ownership of any proposed method, framework, or terminology. Most of the content has been referenced, credited, or taken from public commons. With that, I encourage you to share this content with anyone who might find it valuable.

Format of the Book

This book contains two types of content. The first is a guide to the various methods of Futures Thinking. The second highlights stories from the field, gathered from people I've interviewed who have practiced or are currently practicing some form of futures across both large and small organizations.

METHODS

As a designer, facilitator, and strategist, I've always been fascinated with methods and frameworks. They are the tools of our trade. Coupled with rigorous research, well-organized planning, and a curious imagination, these tools can be effective vehicles for design, alignment, prioritization, and creativity. They are also a way to bring peoples' thoughts and perspectives together to solve problems and devise new and innovative solutions. This is why a major section of the book is dedicated to the HOW: how does futures work, and how can we make it work for us through its methodologies? The description of each method includes an outline of its benefits and cautions to provide varying points of view, as well as caveats for how you might (or might not) use it. The time it takes to conduct each method will vary depending on the context and participants and how you design it into your timeline or agenda. I'll also describe how you might mix, match, or modify certain tools to accomplish different goals. Each method has its own strengths and weaknesses, and I hope that as you read through the methods you can form an objective point of view about how best to use them for your own needs.

STORIES FROM THE FIELD

Throughout each section you'll encounter quotes and stories from veteran and new practitioners. I've interviewed designers, futurists, and strategists to collect their stories of success or failure and have positioned those as examples of what to do or *not* to do. My opinions here are my own, and you can translate what you want from my words, but my overall intent is to highlight different approaches and discuss various ways you can use futures in your design and strategy work today. Hopefully these stories inspire you, and you find a way to use them to build your own vocabulary for how to make futures work for you.

Futures Thinking is an evolving practice, and you will learn that there is no single way to use it. Many people are still investigating and applying futures in a number of novel situations and contexts. We live in an exciting time in which we are witnessing great leaps in innovation, but at the same time we face some of the greatest threats to our species and our environment. I hope this book not only clarifies the futures process for you but also empowers you to use it in new and different ways, challenge the status quo, and transform Futures Thinking into a field that can be accessed and utilized by everyone.

Terminology

There are a few terms that you'll see regularly throughout the book. They are used interchangeably, both formally and informally, to abbreviate and achieve more efficiency in the narrative or discussion. I've included a glossary of terminology at the book's end, but here are some terms you'll encounter more frequently:

Design
> When capitalized, Design refers to the broader discipline or industry of design—as opposed to *design* (lowercase), which is used more informally as a verb or noun.

discursive
> This is the adjective form of *discourse,* which is the discussion of or conversation over a topic, or an extended expression of thought on a subject. We use this term to describe work that is either designed for a conversation or meant to provoke a reaction or debate.

foresight
> This is an abbreviated and informal way to refer to the field of Strategic Foresight.

futures
> This is an abbreviated way to refer to Futures Thinking in general or to work that uses Futures Thinking principles, theories, philosophies, or methodologies.

futures work
> Refers generally to work that employs some type of Futures Thinking methods or principles. It could refer to trend or scenario work or to a range of outputs, including future visions/products, strategic roadmaps, workshops, and so on.

methods, frameworks
> Refer to exercises and activities that are used in futures work.

principles
> Refers to concepts, philosophies, frameworks, ideas, or approaches.

vision

A vision of the future can be a product, a service, an initiative, or an aspirational dream. *Vision* is used loosely to describe the "thing" you want (or don't want) in the future—a goal, North Star, or lighthouse.

O'Reilly Online Learning

 For more than 40 years, *O'Reilly Media* has provided technology and business training, knowledge, and insight to help companies succeed.

Our unique network of experts and innovators share their knowledge and expertise through books, articles, and our online learning platform. O'Reilly's online learning platform gives you on-demand access to live training courses, in-depth learning paths, interactive coding environments, and a vast collection of text and video from O'Reilly and 200+ other publishers. For more information, visit *https://oreilly.com*.

How to Contact Us

Please address comments and questions concerning this book to the publisher:

O'Reilly Media, Inc.
1005 Gravenstein Highway North
Sebastopol, CA 95472
800-889-8969 (in the United States or Canada)
707-827-7019 (international or local)
707-829-0104 (fax)
support@oreilly.com
https://www.oreilly.com/about/contact.html

We have a web page for this book, where we list errata, examples, and any additional information. You can access this page at *https://oreil.ly/making-futures-work*.

For news and information about our books and courses, visit *https://oreilly.com*.

Find us on LinkedIn: *https://linkedin.com/company/oreilly-media*
Watch us on YouTube: *https://youtube.com/oreillymedia*

Acknowledgments

My thanks go to Angela Rufino and our O'Reilly technical reviewers: Hiram Aragon, Ellen Chisa, Frances Close, Christy Ennis-Kloote, Kevin Logan, Jens Oliver Meiert, Kévin Meunier, Danila Pellicani, and Torrey (Newcomb) Podmajersky.

Special thanks to Ben Lowdon; Lina Rodriguez; Jessica Boiteau; Mom, Dad, and my family; Enric Bayó and Elisava University; Martin Waehlisch and the UN DPPA Innovation Cell team, Sebastian Shahfari-Ibler, and Richard King.

Thank you to all the practitioners and experts who were interviewed for this book: Julian Bleecker, Mark Bunger, Gabriella Campagna Lanning, Jake Dunagan, Alex Fergnani, Livia Fioretti, Rick Holman, Brian David Johnson, Michael Kenney, Tino Klaehne, Thomas Küber, Ilana Lipsett, Riel Miller, Madoka Ochi, Sarah Owen, Anthony D. Paul, Sebastian Plate, Bernd Riedel, Joe Tankersley, Cecilia Tham, Martin Waehlisch, Andreas Wegner, and Shihan Zhang.

And for the support and guidance: Agathe Acchiardo; Linda Akkarach; Alisan and Katrina Atvur Rau; Gem Barton; Cesare Bottini; California College of the Arts; Jorge Camacho; Stuart Candy; the Cover Club; Jose de la O; Tec de Monterey; Jake Dunagan and Ilana Lipsett at Institute for the Future; Ellery Studios; the ERA production team (Yoli Inácio, Natalia Ibárcena, Renato Tarma); Livia Fioretti; Nick Foster and Julian Bleecker at Near Future Laboratory; Andreas and Thomas at Futur2; the Futures School; Futurity Systems (Cecilia Tham, Mark Bunger, Magda Mojsiejuk); all my mentors and colleagues at GE Digital and GE Aviation; Charlie Hartzell; Kevin Hawkins; Katherine Hing; Nic Holden; James Hurlbut; my IXDA fam; Uma Jayaram; Michael Johnston; Inge Keizer; Zainab Khan; Nicklas Larsen and the Copenhagen Institute for Future Studies; Iris Latour and Cooper; Brenda Laurel; Ryan Leveille; Dan Levy; Michael Logan; Dot Lung; Hyo Yeon, Madoka Ochi, Lucy Wagner, Jennifer Killian (JK), Kurt Winkler, Steven Fisher, Jeff Salazar, and all our design and foresight colleagues at McKinsey Design; Terrence Melvin; Loriana Mitchell; Jason Napolitano; Vu Nguyen; Chris Noessel; N O R M A L S; Helen Maria Nugent; Scott Patterson; Galdino Pedron; Greg Petroff; PRIMER conference speakers, facilitators, and attendees; Juliana Proserpio and the Echos team; Monica Quintana and Mindset Madrid; Leslie Roberts; Lourdes Rodriguez; Vuyisile Sisulu; the Speculative Futures Community; Studio Andthen; Kyoko Takeyama; Joe Tankersly; Carola Thompson; Lillian Tong; and Andres Valencia.

To all my amazing clients and students around the world, thank you for your love, support, advice, patience, and courage.

Acknowledgments

My thanks go to Angela Rufino and our O'Reilly technical reviewers: Hiram Aragon, Ellen Chisa, Frances Close, Christy Ennis-Kloote, Kevin Logan, Jens Oliver Meiert, Kévin Meunier, Danila Pellicani, and Torrey (Newcomb) Podmajersky.

Special thanks to Ben Lowdon; Lina Rodriguez; Jessica Boiteau; Mom, Dad, and my family; Enric Bayó and Elisava University; Martin Waehlisch and the UN DPPA Innovation Cell team, Sebastian Shahfari-Ibler, and Richard King.

Thank you to all the practitioners and experts who were interviewed for this book: Julian Bleecker, Mark Bunger, Gabriella Campagna Lanning, Jake Dunagan, Alex Fergnani, Livia Fioretti, Rick Holman, Brian David Johnson, Michael Kenney, Tino Klaehne, Thomas Küber, Ilana Lipsett, Riel Miller, Madoka Ochi, Sarah Owen, Anthony D. Paul, Sebastian Plate, Bernd Riedel, Joe Tankersley, Cecilia Tham, Martin Waehlisch, Andreas Wegner, and Shihan Zhang.

And for the support and guidance: Agathe Acchiardo; Linda Akkarach; Alisan and Katrina Atvur Rau; Gem Barton; Cesare Bottini; California College of the Arts; Jorge Camacho; Stuart Candy; the Cover Club; Jose de la O; Tec de Monterey; Jake Dunagan and Ilana Lipsett at Institute for the Future; Ellery Studios; the ERA production team (Yoli Inácio, Natalia Ibárcena, Renato Tarma); Livia Fioretti; Nick Foster and Julian Bleecker at Near Future Laboratory; Andreas and Thomas at Futur2; the Futures School; Futurity Systems (Cecilia Tham, Mark Bunger, Magda Mojsiejuk); all my mentors and colleagues at GE Digital and GE Aviation; Charlie Hartzell; Kevin Hawkins; Katherine Hing; Nic Holden; James Hurlbut; my IXDA fam; Uma Jayaram; Michael Johnston; Inge Keizer; Zainab Khan; Nicklas Larsen and the Copenhagen Institute for Future Studies; Iris Latour and Cooper; Brenda Laurel; Ryan Leveille; Dan Levy; Michael Logan; Dot Lung; Hyo Yeon, Madoka Ochi, Lucy Wagner, Jennifer Killian (JK), Kurt Winkler, Steven Fisher, Jeff Salazar, and all our design and foresight colleagues at McKinsey Design; Terrence Melvin; Loriana Mitchell; Jason Napolitano; Vu Nguyen; Chris Noessel; N O R M A L S; Helen Maria Nugent; Scott Patterson; Galdino Pedron; Greg Petroff; PRIMER conference speakers, facilitators, and attendees; Juliana Proserpio and the Echos team; Monica Quintana and Mindset Madrid; Leslie Roberts; Lourdes Rodriguez; Vuyisile Sisulu; the Speculative Futures Community; Studio Andthen; Kyoko Takeyama; Joe Tankersly; Carola Thompson; Lillian Tong; and Andres Valencia.

To all my amazing clients and students around the world, thank you for your love, support, advice, patience, and courage.

A Fascination with the Future

Before diving into process and methods, I typically like to introduce the concept of *futuring* (the activity of discussing, speculating, and trying to interpret the future) by looking at how we as humans have tried to interpret, predict, and manage future events for exploration or survival. This is both a way to deconstruct how future-tellers have operated throughout time and a metaphorical device for describing these innate abilities, as well as a way to compare futuring to other systems that we've designed for navigating uncertainty. Our species has always been fascinated with the future and with anything that could impact our fate on this planet, and for thousands of years we've looked to those who have been ordained with special gifts or skills to reveal their premonitions to us. In the distant past, prophets and soothsayers were futurists who were seen as *fortune tellers*—gifted individuals endowed with the gift of clairvoyance. Prophets could speak to the gods and had visions that could determine certain events. Though their methods weren't necessarily based on what we would today consider *scientific* principles, these oracles were highly regarded as consultants and confidants to leaders or governments; they had the ability to explain certain phenomena or sway the tides of war or political decisions in a certain direction. Fundamentally, this was an early form of futuring, as these were people who used patterns (and sometimes interpretations of dreams and the imagination) to describe what they saw, which is not too dissimilar to what we do today.

But even before these specialized roles existed in society, we as individuals had been using various devices, models, maps, and stories to describe the future and the unknown. Whether we realize it or not, we are obsessed with the future, largely for reasons of survival, curiosity, and progress. We speculate about the future every day. Whether we're thinking about a trip to the store, planning a vacation, mapping out a career path, or trying to figure out how humans might fly to Mars, forming conjectures about the future is an inherent ability; and those mechanisms are, in fact, the same kinds of tools we use in Futures Thinking (we just put different frameworks and labels around them). We absorb what we know about the world around us and make connections and assumptions about consequences so we can figure out how to attain our goals.

The world has recently survived a pandemic and continues to experience wars, economic crises, and the effects of climate change, and thus there has been renewed interest in futures and strategic thinking so that we can look at more societal, ethical, and globally impactful issues such as environmental crises and political instabilities. But there is still much that is uncertain. In 2002, US secretary of state Donald Rumsfeld, referring to the investigation into weapons of mass destruction in Iraq, said at a press conference (*https://oreil.ly/ju5So*) that "there are known knowns; there are things we know we know. We also know there are known unknowns; that is to say we know there are some things we do not know. But there are also unknown unknowns—the ones we don't know we don't know." This statement of possible unknowns eventually became known as the Rumsfeld Matrix[1] (see Figure 1-1) and is an example of the different types of uncertainty that can exist.

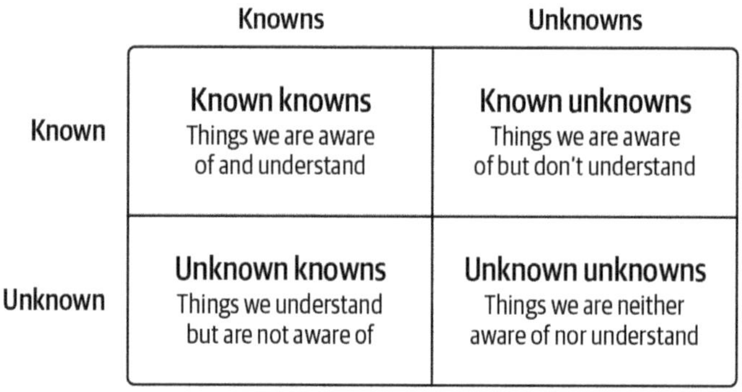

	Knowns	Unknowns
Known	**Known knowns** Things we are aware of and understand	**Known unknowns** Things we are aware of but don't understand
Unknown	**Unknown knowns** Things we understand but are not aware of	**Unknown unknowns** Things we are neither aware of nor understand

Figure 1-1. The Rumsfeld Matrix

The Rumsfeld Matrix has four quadrants (Figure 1-2):

---

1 Also known as the Johari window, a tool developed in 1955 by the American psychologists Joseph Luft and Harrington Ingham to help people improve their self-awareness and interpersonal relationships. It is used frequently in both counseling and corporate team development settings.

Known knowns

These are facts or variables that we're aware of and understand. They form the basis of our knowledge and provide a solid foundation for decision making.

Known unknowns

These are factors we know exist but don't fully understand. They represent gaps in our knowledge that we must address through research, investigation, or consultation with experts.

Unknown knowns

These are elements that we don't realize we know. They're typically buried in our subconscious, overlooked, or dismissed as irrelevant. Uncovering these insights can lead to surprising breakthroughs in decision making.

Unknown unknowns

These are factors that we're not aware of and can't predict. They represent the most significant source of uncertainty and risk, as they can catch us off guard and derail our plans.

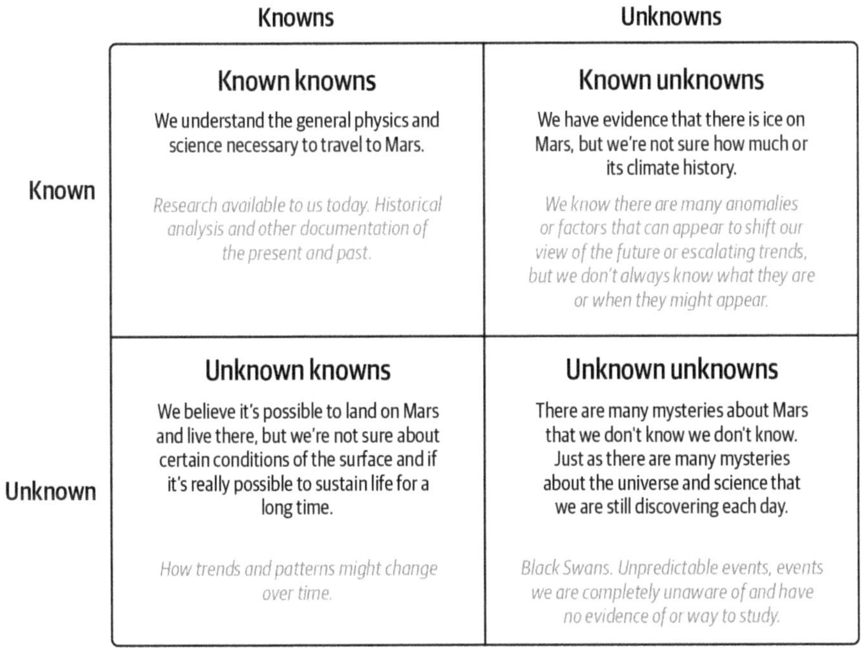

	Knowns	Unknowns
Known	**Known knowns** We understand the general physics and science necessary to travel to Mars. *Research available to us today. Historical analysis and other documentation of the present and past.*	**Known unknowns** We have evidence that there is ice on Mars, but we're not sure how much or its climate history. *We know there are many anomalies or factors that can appear to shift our view of the future or escalating trends, but we don't always know what they are or when they might appear.*
Unknown	**Unknown knowns** We believe it's possible to land on Mars and live there, but we're not sure about certain conditions of the surface and if it's really possible to sustain life for a long time. *How trends and patterns might change over time.*	**Unknown unknowns** There are many mysteries about Mars that we don't know we don't know. Just as there are many mysteries about the universe and science that we are still discovering each day. *Black Swans. Unpredictable events, events we are completely unaware of and have no evidence of or way to study.*

Figure 1-2. An example of a Rumsfeld Matrix regarding travel to and colonization of Mars

Even with all the sensors, data, and algorithms we use to try to understand the future, there are always unknowns, and there will always be some level of speculation. However, *speculation*, which is a facet of futuring, doesn't have to be a wild guess without boundaries or evidence. The word itself seems to suggest something that could be unreliable or unbelievable. But we speculate about things every day; some things we have a little more evidence to speculate about than others, but essentially there are always some levels of uncertainty. And even defining those types of uncertainty based on what we know is still somewhat of a guess. But we continue to do it anyway.

If you are going to walk to the store and the weather report says it's going to be cloudy, it might be so. But it could also rain. And potentially the weather predictions don't account for the small pocket of rain that just might interrupt your path to the store (places like New York and San Francisco have microclimates that are quite unpredictable). The rain is an uncertainty you may or may not have planned for, but you go out anyway. If you are caught in the rain without an umbrella, you get wet. If you thought about that uncertainty and brought an umbrella, you stay dry. Our path into the future is similar. We have to set sail into it because we have a mission to get somewhere. It might rain on us, or it might not. We can try to be prepared by bringing an umbrella in the event that that uncertainty comes true. But this also requires a belief system and ultimately a leap of faith. But then you might ask: if there are so many uncertainties, why do we even bother trying to understand the future? We could ask the same question of our ancestors. Humans, like all animals, exist in complex, volatile, and uncertain environments, and we must rely on our senses and an ability to comprehend and predict patterns around us so that we can successfully navigate the world and the future. This is one of the fundamental concepts of Futures Thinking.

Fail we may, sail we must—Dan Levy, an innovation and design leader, and founder and principal of More Space for Light, has this statement immortalized in a tattoo on his arm. It serves to remind him of his purpose in his work and in all aspects of his life. Failure in business and in life is unavoidable. The only way to make progress is to accept this as a fact and keep moving forward. Levy helps his clients embrace this attitude for strategizing on new and potential futures.

In this chapter I'll walk through several examples of how humans have tried to predict and imagine the future—from our early cave-dwelling ancestors to the science fiction writers and filmmakers of today. I'll also discuss some analogies for how futuring is really just a system of mapping out the world around us and ahead of us, a way to safely navigate through time and space, and a way to strategize about consequences and implications (both positive and negative). By connecting history and the natural inclinations that we use to think about the world, we can use these same analogies to make it easier for others to comprehend the process and ultimately use that information to design the future.

Survival of the Fittest

As one of many species trying to survive and evolve on the planet, humans have found numerous ways, whether through images or storytelling, to document the past and present as a method of recordkeeping for future generations. Some of what our ancestors memorialized in caves or in ancient buildings was left behind intentionally as a guideline for survival; some of it they created merely to capture life as it unfolded around them. Some of the earliest records of these observations go back more than 17,000 years with the paintings found in the caves of Lascaux (Figure 1-3). Though it's difficult to truly explain the meaning of some of the images, some theorize that they are a documentation of past hunting successes, or that they could be tied to rituals (*https://oreil.ly/sccdR*) or stories to be passed on to future generations as cautionary tales, or as strategic planning, or to invoke wonder at and appreciation for the beauty and dangers presented by the world.

Essentially, these were images that the artists reproduced from common patterns in their environment. The logic of observing and predicting those patterns came with a necessity to feed, migrate, or teach. But we aren't the only animal that observes patterns to anticipate the future. In 2004, flamingos, elephants, and cats in Japan were seen behaving erratically and trying to quickly migrate to higher ground (*https://oreil.ly/R1HYH*). This behavior was observed and recorded as many as six days before a magnitude 9 earthquake hit off the coast of northern Sumatra and sent a massive tsunami onto the shores of over a dozen countries, killing more than 225,000 people. Scientists believe that certain animals have incredibly acute senses that might enable them to hear or feel the earth's vibrations, tipping them off to approaching disaster long before humans realize what's going on. While these animals aren't necessarily reading market trends, they are using their natural sensors to listen to signals of changes in the environment. Sensing changes in the vibrations of the earth is no different than monitoring "vibrations" in the stock market. With the help of artificial intelligence, we will soon be able to develop acute senses about the world around us. With inputs coming from multiple sources (social media, videos, news), it is possible that AI will be able to accurately predict the future as well as give us suggestions about what to do should certain scenarios materialize. With the advent of nanotech, wearables, and other devices that can connect our bodies and brains to this kind of intelligence, we could become a new type of tech-enabled fortune tellers.

Figure 1-3. Cave paintings of Lascaux, France (source: subarcticmike)

Meteorology

There are Indian manuscripts from as far back as 6000 BC that tried to predict the weather by observing cloud formations. The ancient Greek philosopher Thales of Miletus has been called the world's first meteorologist; he is said to have predicted the weather and a solar eclipse. And though the weather was sometimes attributed to the work of the gods, some used simple observation and documentation of patterns and commonalities they saw as a logical formula for predicting the weather. This eventually led to what we now call *atmospheric sciences*, a critical field that many agencies (such as those in aviation and the

airlines industry) use every day to plan out any journey on the planet. Today's modern weather forecasting relies on an intricate and orchestrated array of sensors, satellites, and redundancies that come together to analyze and predict what the weather might do. I say *might* because, even with all the satellites and algorithms we've developed, we still are not as accurate and precise as we would like to be at times. Natural disasters such as hurricanes and floods still occur, devastating whole regions, and even after they arrive, we are still guessing where, when, and how they might transform or happen again. Many attempts are being made to try and manipulate the weather through geoengineering (*https://oreil.ly/qUSxP*), an approach that involves projects such as seeding clouds with additives to increase rain or snow, or blocking the sun to reduce harmful radiation from space. But no matter how much we try to predict or control weather events, there is always a force much stronger and more unpredictable than what humans have the ability to control (the unknown unknowns). It's much the same with trying to predict the future. Our instruments can see only so far ahead, and the further into the future we try to look, the blurrier and more uncertain it can be. We can hope and pray that we are right, but we will never be able to fully control certain events or how our actions will impact the emergence or transformation of certain events. Yet we still try, and that is also a fundamental principle of thinking about the future: no matter what the data says (good or bad), we still must try to understand it to avoid certain risks or facilitate certain opportunities.

Mapping the Unknown

Conceptually, maps have always been more than navigation devices; they're a way for us to find our way in the dark, avoid obstacles, and find safe passage to our destination. They have served as both a documentation of landscapes and a tool for current and future generations to move about safely and trade or transport food and materials. But due to the limitations of technologies through the eras, there have always been uncharted territories that we have been forced to speculate about. Even today, with all our fine instrumentation, there is still much of the galaxy (not to mention the universe) that we know nothing about, and we have to use triangulations of data to understand what is out there and how we might survive were we ever to travel beyond the earth's moon. I like to think about futuring as if we are drawing a map into the unknown. There's information that we know to be true (known knowns) and that will continue to be true based on evidence, and there's information that we have to speculate about (known unknowns, unknown unknowns). Drawing a map allows us to identify obstacles and channels that we want to travel within, explore, or avoid completely.

Some of the earliest maps were based purely on memory and personal accounts as well as on mythologies derived from spiritual or religious stories. The oldest known map, the Babylonian *Imago Mundi* (*https://oreil.ly/2nHCG*), is a map of the Mesopotamian world that dates back to the sixth century BC (Figure 1-4). Consisting of simple geometric shapes with cuneiform inscriptions, it places Babylon, an ancient city once located in southern Iraq, in the center, surrounded by what seem to be mountains and bodies of water that likely represent the Euphrates River and the sea. The map is divided into eight regions, with descriptions such as "where a horned bull dwells and attacks the newcomer," "where the morning dawns," "where one sees nothing," and "the sun is not visible." Though parts of the original tablet map are missing, it gives us a peek into the vocabulary and stories that were used to describe unknown regions at the time and the cultural values that existed back then.

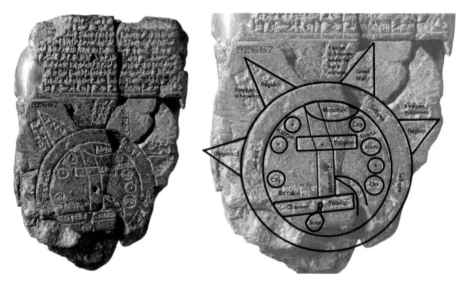

Figure 1-4. The Imago Mundi, *from circa sixth century BC (source: Wikipedia)*

Maps, much like futuring, are a way of understanding what is around us and where to go to avoid danger. As we began to explore and discover new parts of the world in the centuries after the creation of the *Imago Mundi*, we developed more advanced mapping techniques, eventually giving rise to modern cartography (*https://oreil.ly/53i-l*). In addition to personal accounts or drawings, the migration patterns of birds and animals as well as the locations of stars were used to triangulate and detail areas of the world. After the discovery of the Americas, Juan de la Cosa, a Spanish cartographer, explorer, and conquistador, created his *Mappa Mundi* in 1500; this was one of the first recorded maps to include the New World (Figure 1-5). Though there was still much that was unknown and undocumented about what lay on the other side of the planet, de la Cosa tried to fill in the gaps by interpreting what information he did have. On the right side of the map, much of Europe, Africa, and Asia are populated and detailed, while the left side is an inaccurate landmass devoid of much civilization but rich with green mountains and valleys. The accounts of explorers who had made it back from the New World became the only input to inform the map's design. Sometimes, even today, we must rely on experts or other voices to fill in gaps about the future. We may not have empirical evidence to support certain ideas or assumptions, but we have to trust them and try to describe what the world could look like. The lack of precise data regarding a future occurrence has never stopped us from documenting what we think is out there and following it until we are proven wrong. And even then we are able to modify our understanding based on what we see occurring or changing in the world.

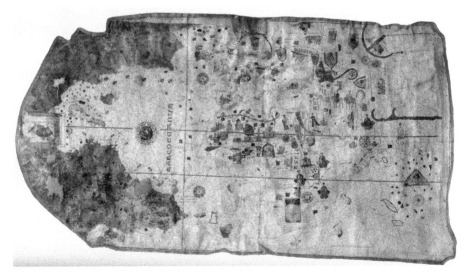

Figure 1-5. Map by Juan de la Cosa showing the Old World and the New World (1500)

In our heads we are constantly mapping out the future, thinking about different scenarios, and trying to predict what is ahead and how we might navigate turbulent waters or obstacles we encounter. (In Chapter 8 we'll look at a method called the Futures Wheel, a diagramming tool used to map the implications of events, trends, or ideas that is, in turn, a way of mapping the future.) But even with a visual representation of the world around us, there are still many factors that could impede our journeys. In the past, a sailing voyage across the Atlantic could be riddled with uncertainty and unknowns, including the weather, human hardships, the complexities of engineering related to sailing large vessels, and even the unpredictable behavior of crews that had to endure such long journeys. To successfully traverse the ocean from east to west, much more planning had to be done, considering it could take several weeks or months just to reach a destination. Thus maps may give us a compass or directional guidance, but there are always uncertainties that we have to prepare for.

As we dive into the stages and activities of Futures Thinking, consider that you are on a fact-finding mission to create your map to a foreign land. The research we do will inform that map and will eventually be used to guide us into the future, avoid obstacles, and create a safe path for us to succeed. It won't always be 100% accurate, but if we do the work, we can gain a much clearer understanding of where we want to go and what we want to do when we get there. But just like traditional maps, our maps of the future must be continuously updated as we receive more information over time.

Imagining the Future

Even some of the early religious prophets would use stories as a way to foretell future events. They were not only orators but also charismatic storytellers, because it was the drama and delivery of their stories that would help create the impact that would inspire or terrify those who sought information. Today, psychics and fortune tellers are seen more as entertainers and are known to use techniques to home in on the sensitivities and body language of audience members in order to pick up clues and deliver predictions about their lives. The story, in turn, is just as important as the future that it portrays. Neither magicians nor psychics have real "magical" abilities; instead, they have tactics for eliciting information by monitoring people's responses to very specific questions or prompts. These become the signals that they interpret and generalize from to make their audience believe what they say is relative or true. I'm not saying that futurists are deception artists like psychics or magicians, but we do use certain methods to try to pick up on clues in the world and interpret what we want to know about the future. And we in turn transform that information into sometimes dramatic stories that can incite change or transformation. But part of our job is also to fill in the gaps. And some of those gaps are based on knowledge, experience, and or even pure assumption.

In a 2021 article in *Wired* (*https://oreil.ly/ehvxN*), Amanda Rees wrote:

> *Since the earliest civilizations, the most important distinction in [how predictions were made and interpreted] has been between individuals who have an intrinsic gift or ability to predict the future, and systems that provide rules for calculating futures. The predictions of oracles, shamans, and prophets, for example, depended on the capacity of these individuals to access other planes of being and receive divine inspiration. Strategies of divination such as astrology, palmistry, numerology, and Tarot, however, depend on the practitioner's mastery of a complex theoretical rule-based (and sometimes highly mathematical) system, and their ability to interpret and apply it to particular cases.*

While words can be an effective way to depict a narrative about future worlds, they are always stronger when paired with a visual representation. Maps are visualizations of the physical world just as much as a graphic novel or comic book is a visualization of a fantasy world, or a designed prototype is a fragment of the future pulled into today. They are no different in function, except they differ in content, intent, and use. Any story, image, or experience about the future is essentially a diagram of that world wrapped inside a colorful frame. We can use such visual diagrams to place people into that world for a moment so that they can experience and react to what it might be like to live there. Just as we use film, literature, and comics to suspend reality so that we can journey into a different time and place, futures work can be much more effective when we illustrate that story (whether rendered statically in an image or distributed into a storyboard or a time-based medium such as film).

In the late 1800s, a French toymaker commissioned Jean-Marc Côté and several other artists to depict visions of the year 2000 (more than one hundred years into their future) (*https://oreil.ly/t-7OS*). Their illustrations (shown in Figures 1-6, 1-7, and 1-8), which originally were in the form of postcards and paper cards enclosed in cigarette and cigar boxes, were distributed between 1889 and 1910 and portrayed a variety of lifestyle scenes and technological advancements in transportation, communication, and sporting activities. The series included at least 87 unique cards. The cards were not made widely available at the time due to financial constraints, but the science fiction writer Isaac Asimov rediscovered them decades later and published them in a book titled *Futuredays: A Nineteenth-Century Vision of the Year 2000* (Henry Holt).

Aero-Cab Station

Figure 1-6. Jean-Marc Côté illustration of flying cars in the year 2000

At School

Figure 1-7. Jean-Marc Côté illustration of a school in the year 2000

A Race in the Pacific

Figure 1-8. Jean-Marc Côté illustration of an underwater race in the Pacific Ocean in the year 2000

In 1930, Echte Wagner, a German margarine manufacturer, produced a line of card sets that were intended for collection in an album (*https://oreil.ly/izr6W*). The album included a section called *Zukunftsfantasien* (Imaginings of the Future) (*https://oreil.ly/MxU1i*). Unfortunately, no artists or authors are credited, but it depicts several scenes of future transportation innovations (see Figures 1-9 through 1-13). While these images illustrate several ideas that have yet to materialize, they also depict some concepts that did eventually become reality, such as wireless phones, televisions, and superhighways.

Figure 1-9. Cover of Echte Wagner Margarine Album Nr. 2

Figure 1-10. Echte Wagner, "Wireless Private Phone and Television." Translation of the verso: "Each person has their own transmitter and receiver and can communicate with friends and relatives using certain wavelengths. But television technology has become so advanced that people can talk and watch their friends in real time. The transmitter and receiver are no longer bound to the location but are carried in a box the size of a photo apparatus." (Source: Klaus Buergle, "Real Wagner Margarine" scrapbooks/Atomic Scout/verso adapted from the German by Otto Z. Mann.)

Figure 1-11. Echte Wagner, "New Highways." Translation of the verso: "The horses are gone, and electricity has replaced the steam power. The pedestrians are no longer in danger from traffic because the motorways and sidewalks are strictly separated. All men and women wear uniform clothing: zipped suits and pants." (Source: Klaus Buergle, "Real Wagner Margarine" scrapbooks/Atomic Scout/verso adapted from the German by Otto Z. Mann.)

Figure 1-12. Echte Wagner, "The Rocket Plane." Translation of the verso: "The aircraft of the future is powered by rockets. The rockets are fitted at the stern of the vessel, which propels the aircraft forward through the recoil of the escaping gases. The aircraft shown here is cruising toward Nankoupas and the ancient Great Wall of China with 10,000 kilograms of mail. Since it has a speed of 1,000 km per hour, it takes less than 8 hours for the Berlin–Tokyo route. A steamer today needs about 50 days!" (Source: Klaus Buergle, "Real Wagner Margarine" scrapbooks/Atomic Scout/verso adapted from German by Otto Z. Mann.)

Figure 1-13. Echte Wagner, "Spaceship Port." Translation of the verso: "Because there are rare minerals on the Moon, America has built a $20 billion enterprise named MoMA-A.G. (Moon Minerals A.G.). At this dock station, the ships can renew their rocket fuel. The station floats freely in space." (Source: Klaus Buergle, "Real Wagner Margarine" scrapbooks/Atomic Scout/ verso adapted from German by Otto Z. Mann.)

In these frozen frames of future worlds, a sequential story is not necessarily being told, but in each image is embedded an implicit narrative about what *could be*, given certain advancements. The artists took scenes of their current reality and transported them into a new one augmented with innovations that provoke new ideas about where we could go if we had the ability. These "visions" may not have been the work of trained prophets or scientists with special abilities, but they are an amalgamation of the artists' own imaginations, dreams, and observations about the world around them, including what they were exposed to and inspired by at the time.

Literature, film, and television have always played a major role in entertainment and inspiration when it comes to imagining future worlds. From the 1960s television series *Star Trek* to *Star Wars* and beyond, imaginations have run wild, mixing fantasy and science fiction to explore new territories. *Star Trek*'s stories take place in 2265 and follow the journey of the starship USS *Enterprise* as its crew discovers new planets and life forms and navigates both military and ethical conflicts. The first episode, which debuted in 1966, introduced many concepts into science fiction that would become inspiration for real technologies to come. One of those concepts/devices was the tricorder (*https://oreil.ly/ztgVr*) (Figures 1-14 and 1-15). The show's creator, Gene Roddenberry, described this device in the series' *Writers/Directors Guide:*[2]

> *TRICORDER: A portable sensor-computer-recorder, about the size of a large rectangular handbag, carried by an over-the-shoulder strap. A remarkable miniaturized device, it can be used to analyze and keep records of almost any type of data on planet surfaces, plus sensing or identifying various objects. It can also give the age of an artifact, the composition of alien life, and so on. The tricorder can be carried by Uhura (as communications officer, she often maintains records of what is going on), by the female yeoman in a story, or by Mister Spock, of course, as a portable scientific tool. It can also be identified as a "medical tricorder" and carried by Dr. McCoy.*

---

2 Reprinted in Paula M. Block, *Star Trek: The Original Series 365*, with Terry J. Erdmann (New York: Abrams, 2010).

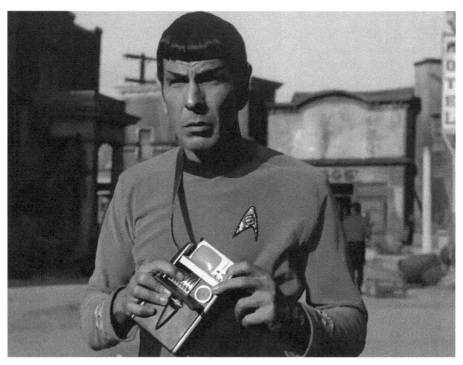

Figure 1-14. Mr. Spock holds a tricorder in the 1966 Star Trek *episode "The Man Trap"*

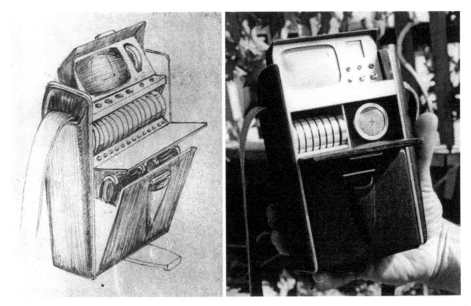

Figure 1-15. An original sketch of the first tricorder designed by Wah Ming Chang for Star Trek *(left), and a prototype used in the series (right)*

The concept of a portable scanning device that can capture a variety of data inspired the Canadian company Vital Technologies Corporation to develop and produce the first "real world" tricorder (*https://oreil.ly/nMkTZ*) in 1996. The scanner was called the TR-107 Mark 1 (Figure 1-16), and it could scan and measure electromagnetic radiation, temperature, and barometric pressure. Vital Technologies sold ten thousand of the devices before going out of business in 1997.

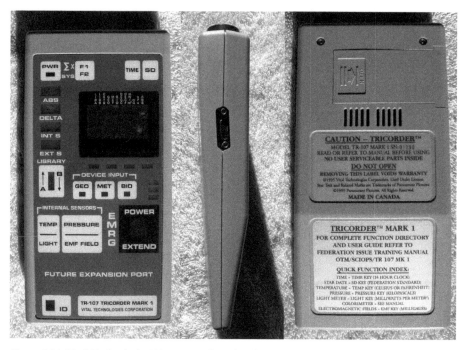

Figure 1-16. Vital Technologies' TR-107 Mark 1 tricorder

In the decades since the show's inception, *Star Trek* has inspired millions of people with its stories that detail the trials and tribulations of exploring unknown worlds while highlighting challenges similar to those we face here on earth with regard to politics, cultures, ethics, and war. And the dream of creating a modern-day tricorder, or a device that can measure many vital statistics about an environment or a human, continues to the present day. Today, the modern Apple Watch can measure everything from your blood pressure to the timing and strength of the electrical signals of your heart, providing doctors with insight about potential heart irregularities and conditions. In January 2024, at the annual Consumer Electronics Show in Las Vegas, the health tech company

Withings debuted a new product called the BeamO (*https://oreil.ly/FdsUS*) that combines four medical tools: a stethoscope, an oximeter, a one-lead ECG, and a thermometer (Figure 1-17).

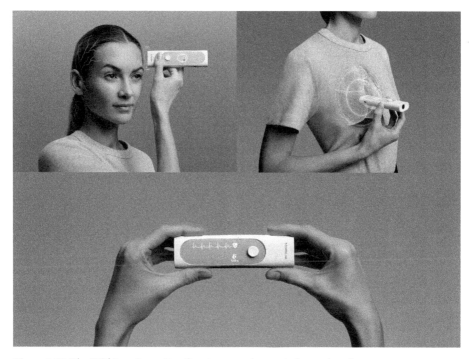

Figure 1-17. The Withings BeamO, a four-in-one at-home vitals monitor that can measure body temperature and blood oxygen levels and that also features a digital stethoscope and a medical-grade ECG2

However, some of the storytelling devices in science fiction are inherently used to create drama and conflict in order to keep readers' attention and create a feeling of excitement or dread around the possibilities of what technology and society could become. Dystopian tales of collapsing societies and of unchecked and uncontrollable technologies are common themes in the sci-fi portfolios of literature and film. Not all science fiction stories are dystopic, however; sci-fi has also been used to explore the optimistic potential of society, envisioning futuristic technologies and the benefits they could have for mankind. Philip K. Dick (*https://oreil.ly/iEVB1*), who wrote dozens of novels and more than 120 short stories, was a popular science fiction author whose stories have been turned into feature films such as *Blade Runner, Total Recall,* and *Minority Report.* Many of his stories were huge successes and inspired countless other science and science fiction authors. The movie *Blade Runner* (Figure 1-18), adapted from his novel *Do Androids Dream of Electric Sheep?*, is a dark vision of a polluted and overpopulated Los Angeles in the year 2019, when the world is populated with genetically bioengineered "replicants" (androids) that are visually indistinguishable from humans. The story focuses on an android-hunting detective named Rick Deckard, who reluctantly agrees to take one last assignment to hunt down a group of recently escaped replicants. During his investigations, Deckard meets Rachael, an advanced experimental replicant who causes him to question his attitude toward replicants and what it means to be human. Laden with flying cars, robots, and futuristic devices, homes, and cityscapes, *Blade Runner* has become a common reference point for anyone contemplating dystopian futures and the potential consequences of artificial intelligence gone astray.

Figure 1-18. The Los Angeles skyline in 2019, as envisioned in the 1982 film Blade Runner

Around 1993, Glen Kaiser, a former product manager at AT&T, a major telecommunications company in the United States, teamed up with David Fincher (later the director of such films as *Fight Club* and *The Social Network*) to develop a marketing campaign and series of television commercials titled "You Will" (*https://oreil.ly/oOPpm*). They created several commercials (Figures 1-19 through 1-22) that showcased emerging technologies and platforms being developed at Bell Labs and AT&T at the time. The narration for the first ad, delivered by Tom

Selleck, began, "Have you ever borrowed a book from thousands of miles away? Crossed the country without stopping for directions? Or sent someone a fax from the beach? You will. And the company that will bring it to you is AT&T." In an interview on Quora (*https://oreil.ly/Alv_r*), Kaiser explained the history behind the campaign:

> *A young team of six people, including me, were charged with developing leading-edge new business services (as opposed to consumer services). I was Product Manager for the Picturephone Meeting Service, one of the key areas where we saw AT&T having a dominant future. We at Corporate did all the research in specific markets of video conferencing, voicemail, messaging, security, and even RFID tracking. Financial business cases were developed, and the labs were funded to build the products and services. It was akin to being at Xerox PARC or at Palo Alto Research Park. I visited all of them. We were well on our way to making these applications commercial using the latest Bell Labs research and technology.*
>
> *In the mid-eighties, AT&T was offered the chance to sponsor Spaceship Earth at Disney's EPCOT Center, showcasing some of our applications. When it launched, EPCOT demonstrated the appeal of AT&T's technology vision. Millions of visitors saw it and were excited by its near-term realization.*
>
> *By 1993, the products we had in mind were not quite ready for prime time—the quality they needed to be at for widespread market adoption was not there yet, and affordability would also have been a factor. But in reality, the concepts were good, the core technology was real, and we just needed to wait for Moore's law and Internet adoption to catch up for us to properly commercialize them....*
>
> *The aim of this advertising initiative was clear: To project a more relevant AT&T to the youth demographic, and build enthusiasm for what we had in the pipeline. We picked all of the applications and technology that we thought were both realistic and would create interest for consumers.*

Figure 1-19. Image from a 1993 AT&T "You Will" commercial (https://oreil.ly/1Pw93) portraying smart home capabilities

Figure 1-20. Image from a 1993 AT&T "You Will" commercial that envisions writing on a digital tablet at the beach

Figure 1-21. Image from a 1993 AT&T "You Will" commercial that portrays a student at their desk listening to a teacher through a remote interface

Figure 1-22. Image from a 1993 AT&T "You Will" commercial showing a child using the interface of a smart TV

The campaign, which showcased many technologies that eventually became a reality, seemed at times to accurately predict the future with great precision. Among the ideas depicted in the short vignettes were innovations such as smart homes, digital tablets, remote work, telelearning, radio frequency identification (RFID), smart tolls, digital medical records, and smartwatches.

In December 2011, Charlie Brooker launched a TV series in the United Kingdom called *Black Mirror*. Each episode has a different plot, actors, theme, and message and is set in a near-future dystopia, usually focusing on the use and implications of fictional technology. The series is inspired by *The Twilight Zone* and uses the themes of technology and media to comment on contemporary social issues. *Black Mirror* has received critical acclaim for its provocative messages about how we are engaged with technology today, how our culture is changing and potentially threatened by the various consequences of tech, politics, and the impact on the human condition. The show's pilot episode, "The National Anthem," did not feature futuristic sci-fi tech, however; instead, it was a commentary on how society consumes and dramatizes news, information, and politics as entertainment, perpetually fueling a machine of drama and exposing people's lives for political influence and to increase ratings, all at the cost of others' personal tragedies. The episode titled "San Junipero" (Figure 1-23), released in 2016, is a love story set in a future in which people can upload their consciousnesses into a virtual world inhabited by the deceased and the elderly. The episode touches on several trends and technologies that were developed or became more widespread during the 2010s: same-sex relationships, virtual reality, online community platforms (early metaverse (*https://oreil.ly/Qqxya*)), and society's obsession with controlling the aging process, as well as digital afterlife.[3] The story is a testament to the potential of technology to assist in euthanasia and the consideration of ethical concerns, as well as a platform for exposing the underlying taboos around same-sex relationships, as these communities have continued to fight for equal rights and representation in film, media, and society.

---

3 Digital afterlife is associated with the volume of the digital assets and footprint left behind by people in the digital world—what happens with all your data once you die and how does it represent you, who owns it, and what happens to it when there is no physical body to own and manage it.

The story synopsis goes like this:[4]

In 1987, the shy Yorkie meets the outgoing Kelly in a beach resort town named San Junipero. The next week, the pair meet again and have sex. Yorkie struggles to find Kelly afterwards, until a man suggests looking in a different time. She searches in multiple decades until finding Kelly in 2002, where Kelly confesses that she is dying, and wanted to avoid developing feelings for Yorkie. They have sex again. San Junipero is revealed as a simulated reality inhabited by the deceased and the elderly, who interact through their younger bodies. In California, Kelly meets a paralyzed Yorkie, soon to be euthanized so that she can live in San Junipero permanently. Kelly marries Yorkie to authorize the euthanasia. However, the pair argue when Kelly says she does not wish to stay in San Junipero when she dies: her husband, with whom she was together for 49 years, did not choose to join after their daughter died without the option to do so. After some time, Kelly changes her mind and happily reunites with Yorkie after her own euthanasia.

Reviewers have described "San Junipero" as a highly optimistic, emotionally rooted love story and a work of science fiction. It features the first same-sex couple in **Black Mirror**. *Rebecca Nicholson of* **The Guardian** *wrote that it "leaves you believing in the power of love to fight pain and loneliness." Some reviewers noted that the love story "transcends consciousness." The episode also has unhappy elements and has been called "bittersweet."*

---

4 Wikipedia, s.v., "List of *Black Mirror* Episodes" (*https://oreil.ly/nsSm6*), last modified March 24, 2024, and Wikipedia, s.v., "San Junipero" (*https://oreil.ly/AxSP6*), last modified February 17, 2024.

Figure 1-23. A scene from the Black Mirror *episode "San Junipero": Kelly and Yorkie having a drink at Tucker's*

In March 2020, the director Alex Garland released a limited series called *DEVS* on the FX network. (Spoilers ahead!) The series, a science fiction thriller starring Nick Offerman and Sonoya Mizuno, is based around a fictional company called Amaya that eerily echoes tech giants like Google or Facebook. Situated in an area that resembles Silicon Valley, Amaya is a quantum computing giant that has seemingly cornered the market with its technologies (Figure 1-24).[5] But hidden within a bunker in the Amaya compound is a very secretive, highly secure

---

5 Quantum computing is a rapidly emerging technology that harnesses the laws of quantum mechanics to solve problems too complex for classical computers. Unlike classical computers that process information on a series of bits and bytes of binary data (1s or 0s), quantum computers use Qbits, a type of information that can be a 1 and an 0 at the same time, removing the time and energy it takes to read and write a linear string of code of binary data. Quantum computers have already made enormous advancements, such as in the ability to solve mathematical problems in mere seconds that typically were unsolvable by classical computers.

project called DEVS that is protected by a Faraday cage and that only selected elite engineers are allowed to work on. The series centers on the disappearance of an engineer's boyfriend shortly after he begins work on the project. As the engineer (Mizuno) delves deeper into what DEVS really is, she discovers that not only has Amaya been using its quantum technology to predict certain events in the world, but it also has the ability to look back in time thousands of years. The proposition of *DEVS* speaks to much of the cultural fascination and mystery around what quantum computers are capable of. Today only a few companies in the world—Google, IBM, and Microsoft, for example—have quantum computers, and they are being used in the areas of physics, communications, and other mathematical problems that have been deemed too costly or difficult for classical computers to solve. But the promise of quantum is just on the horizon. *DEVS* is a speculation on what we could do with quantum if we had the power to apply it to predicting the future and seeing deep into the past. What are the implications? And how could it change the way we view or navigate the world if the future could be predetermined by a machine? Where does human agency ultimately play a part? But also, what are the dangers associated with wielding such power, and what ethical concerns could arise?

Figure 1-24. Quantum computer in the FX series DEVS *(photo by Miya Mizuno/FX)*

Whether in static images or in the moving images of film and TV, the future has always been illustrated in different formats to draw us into possible worlds so that we can either internally or externally participate in a discourse around "What if?" questions. The statements embedded in some of these examples may vary based on the authors' intended commentary or the era the authors were living in, but they were all created to serve the purpose of describing the future with a sense of wonder and excitement, as well as with caution and deliberation. Not all of these visions came true, but some certainly did. The intent is not to be accurate or correct, but to use them as a platform to fuel discourse, imagination, transformation, and innovation. As technology, media, and the connections of global communities expand, contract, and evolve, there are more opportunities every day to inspire people about the future through storytelling, using simple devices of narratives, visualizations, and drama. One thing is certain, however: we will have no shortage of visions for generations to come.

In Closing

Humans have an innate ability to speculate about and visualize the future. It's in our DNA, a culmination of our ancestors' desire to understand the world and survive within it. As you begin to explore the principles and methods in this book, you'll see that these are just additional tools and devices that can be used to navigate time and space. Regardless of how you are applying these methods (for projects, strategic roadmaps, innovation, or planning your career), the mechanics of these methods shouldn't differ from any subconscious process you use in your head today when you are strategizing how to plan a vacation, map out your career, buy a car, or start a family. Sensing and gathering these signals and vibrations in the world will allow you to easily construct your own map of the future and analyze how it works so that you can sail into and through the future more safely and productively. Once you're able to tap into this ability and internalize the process, it won't feel complex or intimidating. In the end, futuring should be exciting. No matter what threats or barriers you discover, you will hopefully be able to invite these uncertainties into your world as exhilarating problems to be solved.

A Primer on Process and Approach

Before we dive into the principles, processes, and methodologies of Futures Thinking, this chapter provides a bird's-eye view of each of the distinct phases of work and the activities that occur within each step (Figure 2-1). *Making Futures Work* ties together Strategic Foresight, Design Futures, and Strategy as an inter-dependent model for *designing* the future. The phases outlined here are steps that build on one another, ultimately converging on an actionable strategy for bringing a future vision into fruition. As discussed in Chapter 1, when we are trying to understand certain complexities or unknowns, we try to build maps to help us navigate. Thus, this chapter is a map of the book and is meant to prepare you for what's ahead. You can certainly skip this section and dive right in, but you may find it helpful to see how we will progress so that you don't get lost along the way.

Historically, the *craft* of Design hasn't necessarily played an integral part in Strategic Foresight or Futures Studies, though there definitely are artifacts that may be *designed* or illustrated. For the most part, foresight work may rely only on design to visualize a scenario or vision of the future and may not necessarily employ other capabilities that designers use, such as prototyping and user experiences. But design has more to offer across the field of futuring, and this book is a testament to how designers can play an integral role in every phase of the process. Some practitioners have different approaches or may omit a stage or method for their own reasons, and you certainly have the agency to practice however you like, but the intent behind laying out these steps is to demystify the vocabulary and philosophy behind fundamental principles and give you a few practical ways to get started in or continue your exploration of futures work. While these sections are presented as numbered stages, you aren't required to carry out your futures work in this order. I have detailed the process in this way to present a general sequence of events that I've found useful for engaging in futures work so that it leads toward a strategic vision or toward a particular future goal (a designed product, service, initiative, or aspirational goal). The sequence outlined here may not always align with your context or needs, so feel free to use the stages as loose guardrails for the incremental phases of your futures work:

- Stage 0: Preparing for Futures Work
- Stage 1: Strategic Foresight
- Stage 2: Designing Futures
- Stage 3: Designing Strategy

Stage 0 Preparing for Futures Work	**Stage 1** Strategic Foresight	**Stage 2** Designing Futures	**Stage 3** Designing Strategy
Preparing yourself and your client for futures work	Analyzing patterns to build future worlds and scenarios	Designing future initiatives, visions, products, or services	Building a roadmap toward future visions
• Understanding your audience • The futures mindset • Problem framing • Project framing	• Working with trends • Analyzing and prioritizing trends • Scenarios • Implications	• Speculative design • Design fiction • Sci-fi prototyping • Experiential futures	• Backcasting • Measuring success • Culture building • Integrating into your organization

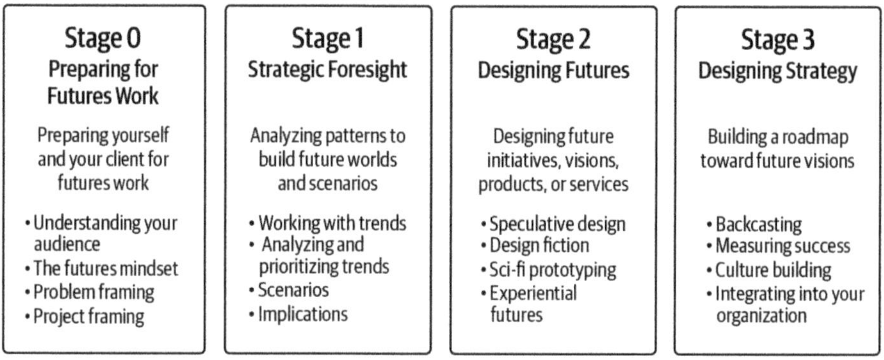

Figure 2-1. Stages of using Futures Thinking for strategy

Stage 0: Preparing for Futures Work

Before engaging in any futures work (or in strategic work in general), it's always important to understand the context and problem you are trying to solve. We never dive in headfirst without knowing what we're diving into. And while you might be excited to learn the methodologies, I implore you to consider this first stage as a necessity in futuring, as it will properly prepare you and your teams for the potentially difficult but exciting journey ahead. Without a proper understanding of your audience, environment, needs, and complexities, you could be caught off guard and face tensions, skepticism, and other challenges that could potentially derail your efforts.

UNDERSTANDING YOUR AUDIENCE

In this chapter I'll discuss the idea of a "diagnostic" as a framework for investigating the culture, people, and context within which you'll be doing futures work. Acquiring a good understanding of who you'll be working with, how they do strategic planning today, the associated complexities and barriers, and what has succeeded or failed will give you a lot of fodder to properly prepare your teams for the journey and help you execute more smoothly. A good diagnostic of the audience also allows you to plan the process, methods, and outcomes more effectively.

THE FUTURES MINDSET

Once you've properly assessed the situation and client culture, you'll have enough information to understand how much you need to prime your client, team, or audience for the journey. Here I'll discuss the idea of a *futures mindset*— a way of thinking, behaving, and operating that allows them to embrace the process and the uncertainties the future holds. A healthy futures mindset will allow everyone to optimistically discuss the threats and opportunities you discover with confidence and excitement. If not carefully managed, teams can easily become overwhelmed by terminology, process, or the sheer volume of possibilities you uncover. Ultimately, nurturing this mindset can sustain everyone's investment throughout so that you can get to the finish line with minimal opposition or attrition.

FRAMING THE PROBLEM

Typically, you never want to jump into a project without outlining what you are doing and where, when, and why you are doing it. Framing these details will be critical for starting your research and providing guardrails as you traverse the activities. This chapter introduces some of the key components for framing the problem for a futures project, including the focal issue (what topic you are interested in), geography (where in the world you are looking at), the time horizon (when in the future you want to explore), and other details that support the investigation you are about to embark on.

FRAMING THE PROJECT

This chapter exists for good housekeeping and stresses the need to be organized and to advocate for your participation in the logistical aspects necessary for undertaking a potentially large body of work. If you are working on a 10-year strategy for a company, it will be important to have healthy communication, stay organized, and maintain a healthy balance among time, funding, roles, and activities. While some projects might have a dedicated project manager, you may be the one wearing many hats and directing the entire project. So whether or not you are leaning on someone to keep you organized, this chapter will provide some guidance on key aspects that will help you plan, strategize, and deliver on the project.

Stage 1: Strategic Foresight

Strategic Foresight (*https://oreil.ly/tZ2j8*) is one of the core disciplines within Futures Thinking and provides an excellent framework for conducting research and building future worlds and scenarios. It's used to gather and process information about future operating environments. This information can, for example, include trends and developments in their political, economic, social, technological, and legal environments. To understand a world that hasn't arrived yet, we have to use a variety of lenses to gather information about the future, analyze and prioritize that information, and project into a future time and space to develop perspectives on how the future could unfold. Historically, the field of Strategic Foresight has found application in diverse settings, ranging from the military to business schools. Its origins trace back to the 1950s with the establishment of the RAND Corporation by the US Air Force in the aftermath of World War II. The cases we'll discuss in these chapters will explain how anticipatory thinking has protected some of the world's largest corporations from economic crises, and how some organizations use it for innovation to develop new policies, processes, and technologies. In essence, foresight captures the work we do as researchers, sociologists, anthropologists, and ethnographers. It involves taking into account history, culture, market trends, political movements, and anything and everything that can contribute to the body of knowledge we need to form a perspective about the future we want to design for.

TRENDS AND FUTURES INTELLIGENCE

Where do we find the information we need to understand the future? We can start by identifying the trends and patterns we see around us today and in our past. Because we don't actually have a way to contact the future, we have to use what we have in the present to make our own assumptions and speculations. We do this every day when we plan vacations or events. However, no matter how much we try to be prepared, anything can happen. And part of the process of scanning the horizon and world for patterns is being aware of those uncertainties and working with them in a way that is manageable and prioritized so that we can tackle the things we care most about. I'll discuss the different types of trends (mega, macro, micro, and fads), where to find them, and other types of futures intelligence we can gather to develop our perspective on the future.

ANALYZING TRENDS

There potentially are mountains of information on trends and other data sources; how do we analyze and prioritize what's most important for us to address? The overwhelming bodies of data you gather will require a method of selecting and decision making so that you can converge on more concrete views of future scenarios and worlds. I'll dive into some basic methods for prioritization that take advantage of qualitative and quantitative data to help you make more informed decisions.

SCENARIOS AND IMPLICATIONS

Futuring is all about thinking in "What if?" scenarios. When we talk about Strategic Foresight, scenario planning is usually one of the first things that comes to mind. It is a core activity and seeks to answer questions such as: What could happen given certain conditions that trends are creating? What are the alternative futures that could play out? And how do we ultimately decide which scenario is the one we should plan for? These are all questions that I'll unpack as we learn about traditional scenario archetypes and variations you can use in different situations. But there are many ways to ask questions about the future and discuss scenarios. Implication mapping or implication diagramming is another way to look at different consequences or impacts. Implications are traditionally generated with a *Futures Wheel*, a diagram framed in terms of consequences or impacts due to a particular event, trend, or idea. We'll learn the different variations of implication mapping and how to use them to build scenarios and inform the ideation of future visions.

Stage 2: Designing Futures

For designers, this is naturally an exciting stage of work and I could have written an entire book just on this phase, but through my experience of learning how futures works, I've found it useful to learn and practice the other disciplines such as Strategic Foresight and traditional business and innovation strategy as a way to facilitate a more structured research and analytical framework for informing futures design. Instead of just saying "there's a future, and here's an idea," we can utilize foresight to truly create the world of the future that we want to live in (or not live in). We can look around, consider the context and environment (social, cultural, economic, technological, or political), and use that input to inspire our ideas for future products and services. But even with a more structured method for doing research, we can't forget about the art and craft of design that is fueled by our experiences and imagination.

TYPES OF FUTURES DESIGN

Over the years, many formats and approaches have emerged, creating an eco-system of methodologies that can be used to experience or visualize futuristic concepts. I'll discuss these different types of design futuring, their labels, their similarities, and their differences in practice and application. If you are just discovering this type of work, you might encounter several terminologies that sound or look similar, including Speculative Design, Critical Design, Design Fiction, Science Fiction Prototyping, Experiential Futures, and Futures Design. And while there definitely are many flavors of designed futures, I will discuss each of these to help delineate their different origins and characteristics and will furthermore try to categorize them by how and when to use them in different contexts or for various intentions and purposes.

DESIGNING THE IDEATION WORKSHOP

The second section of the Designing Futures stage is dedicated to designing the futures ideation workshop. In my experience, this is typically a very challenging activity depending on the team, its experience with creative brainstorming, the quality of output you want, and the time you have to execute the activities. Among the challenges I'll discuss in the chapter is: how do you transport some-one into a future world that doesn't exist yet and get them to understand what makes that future different from (yet similar to) today? The process of generating ideas that are novel, provocative, or truly innovative can also be difficult if the environment, inputs, and parameters are not clear or effective. Thus, this chapter is dedicated to introducing some methods and their mechanics to be used in brainstorming and ideation workshops. With the proper selection of tools and facilitation, this is the moment you can really generate exciting visionary prod-ucts, services, or initiatives of the future.

Stage 3: Designing Strategy

The final stage of strategy ties everything together into an actionable plan that we can execute today. Fundamentally, we will use this chapter to understand what strategy means in terms of Futures Thinking; how it's discussed, designed, and measured; and how it can succeed or fail.

PLANNING FOR STRATEGY

I'll discuss various considerations for how to prepare for strategic meetings, whom to invite, setting up metrics, resource planning and alignment, and some activities to help you develop a roadmap toward a North Star. I'll discuss a traditional roadmap planning method called Backcasting, a framework for looking backward from the future to identify the critical steps, milestones, or goals necessary to arrive at that future vision. I'll also discuss how to segment the roadmap into different stages (short-term and mid-term) so that you can successfully measure and connect near term goals with the long-term vision.

MEASURING SUCCESS

One of the most common debates concerns how to measure futures work. How do you set up clear quantitative metrics for something that can take years to arrive? We'll discuss different types of metrics and indicators to watch including quantitative and qualitative measurements so that you can keep an eye on what is making futures work along the way.

Whether or not you are involved in strategic planning conversations today, this chapter is important for understanding how to make futures more than just a thought experiment. But planning a roadmap also requires nurturing a futures culture that is armed with tools and an excitement to contribute organizationally and operationally, eventually permeating many levels of an organization through processes, education, and collaboration. Thus, this final stage of work will not only be about planning but also implementation and provide tips on how to sustain the work as it is executed.

Integrating and Advocating for Futures in an Organization

After you've learned about the fundamental stages and processes, I'll walk you through how to integrate Futures Thinking into an organization, whether you are an in-house employee or an external consultant. I'll discuss some common challenges I and other practitioners have experienced when trying to begin futures or to implement it into teams. Each entry point has its own nuances, and I'll provide a few practical ways to make futures work for you. Hopefully, this final chapter will provide some useful insights into how you can get started, depending on what your situation might be. But it's placed at the end for a reason—because, to make futures work, you first have to learn *how* it works. So let's begin that journey now.

PLANNING FOR STRATEGY

I'll discuss various considerations for how to prepare for strategic meetings, whom to invite, setting up metrics, resource planning and alignment, and some activities to help you develop a roadmap toward a North Star. I'll discuss a traditional roadmap planning method called Backcasting, a framework for looking backward from the future to identify the critical steps, milestones, or goals necessary to arrive at that future vision. I'll also discuss how to segment the roadmap into different stages (short-term and mid-term) so that you can successfully measure and connect near term goals with the long-term vision.

MEASURING SUCCESS

One of the most common debates concerns how to measure futures work. How do you set up clear quantitative metrics for something that can take years to arrive? We'll discuss different types of metrics and indicators to watch including quantitative and qualitative measurements so that you can keep an eye on what is making futures work along the way.

Whether or not you are involved in strategic planning conversations today, this chapter is important for understanding how to make futures more than just a thought experiment. But planning a roadmap also requires nurturing a futures culture that is armed with tools and an excitement to contribute organizationally and operationally, eventually permeating many levels of an organization through processes, education, and collaboration. Thus, this final stage of work will not only be about planning but also implementation and provide tips on how to sustain the work as it is executed.

Integrating and Advocating for Futures in an Organization

After you've learned about the fundamental stages and processes, I'll walk you through how to integrate Futures Thinking into an organization, whether you are an in-house employee or an external consultant. I'll discuss some common challenges I and other practitioners have experienced when trying to begin futures or to implement it into teams. Each entry point has its own nuances, and I'll provide a few practical ways to make futures work for you. Hopefully, this final chapter will provide some useful insights into how you can get started, depending on what your situation might be. But it's placed at the end for a reason—because, to make futures work, you first have to learn *how* it works. So let's begin that journey now.

Preparing for Futures Work

Stage 0 is the initial stage of doing futures or any other kind of strategy work. It involves the data gathering you'll need to do to prepare for doing futures work. Before beginning any futures project, I've found that it's important to understand the team, client, or organization and the problem space you are going to be exploring. This includes making sure everyone is clear about the principles, stages, and outputs of the process and how it ultimately delivers value. This will allow you to avoid confusion or anxiety down the road as you begin generating a wealth of content and possibilities.

In this stage, we'll cover the following:

- Chapter 3, "Understanding Your Audience"

    — Use situational awareness methods to understand your audience, team, or organization: the projects, the people, the environment you'll be futuring in

    — Determine how you might want to design your project to fit that environment

    — Decide how much you will have to prepare your client or team for the process and journey

- Chapter 4, "The Futures Mindset"

    — Begin preparing them for futures work by showing them examples, explaining what you do and explaining the process

    — Make sure they're ready, aligned, and excited for the work ahead

- Chapter 5, "Framing the Problem"
    - — Determine the detailed parameters of the project: focal issue, time horizon, geographic boundaries, demographics, complexities
- Chapter 6, "Framing the Project"
    - — Get organized to start the project
    - — Set up the team, timeline, deliverables, funding, logistics
    - — Prepare the tools, repositories, and processes

Understanding Your Audience

Futuring is both a mindset and a journey, and as on any journey, it's essential to have a clear understanding of your destination as well as a plan for how you'll successfully get there. Let's use the analogy that you are sailing a boat to an uncharted island. No one has been there before, but you might have some satellite data about it. You know roughly where it is, and you know it is possible to get there on the right kind of vessel. But before you set sail, you have to do a bit of research. What kinds of threats might exist? What can you take advantage of, and what do you need to protect yourself against? It might be a treacherous place but may also be a place of beauty, bustling and full of life. And while you can try to gather as much data about the island as you can through various instruments, there will always be some level of speculation. The future is no different: it's uncharted, and there are always surprises, and how we handle those surprises will determine our fate—our survival or our demise. Similarly, businesses, designs, projects, and strategies can live or die depending on whether we are prepared for what's ahead.

But the island isn't the only thing you have to prepare for. You need to prepare for the journey itself. The trip could take days, and you might be sailing with complete strangers. Do you normally just jump into a boat with strangers and set sail to an uncharted territory? I hope you'd be smarter than that and ask lots of questions to make sure you know what you're getting into. If it's the difference between living or dying at sea, you're probably going to want to know everything there is to know about who you're going with, the boat, the route, safety procedures, the island, and how you will survive the journey. On this trip, you (as the futures leader) are the captain; similarly, you'll need to do some research about your passengers.

In this chapter we'll discuss some traditional methods of inquiry. And since there are many ways to do inquiry, either through interviews or workshops, the idea is to use what you have at your disposal to gather the information you need from your passengers so that you can prepare them for the journey. With these tools, we can understand our client or audience—the issues they have today (their motivations), what outcomes they desire (the expected deliverables)—and analyze the situation (situational awareness) so that we can properly plan and use the process. Some of this guidance may sound like traditional consulting protocol, and indeed, some of it is. But this inquiry should also be used on any kind of engagement when you're starting a project. It's good housekeeping, and a way to set up a healthy start to any journey.

The Diagnostic

In Stage 0, we are preparing ourselves and our audience for the journey ahead. To do this, we want to begin by asking a lot of questions and listening. I like to call this step the *diagnostic*. The word *diagnosis* might be more commonly associated with healthcare or car maintenance, where a professional runs several tests or queries to determine what needs to be done to help you or your vehicle. This doesn't necessarily mean that people or organizations who want to look into the future have a disease or disability. A diagnostic can be seen as a simple query to understand a situation. A futures diagnostic is similar in the sense that it's probably best to understand the characteristics of your team, project, or operation before you make any decisions about *how* to work with them. You begin by investigating who your stakeholders are and their roles, perspectives, working culture, projects, and expectations, as well as what barriers to progress might exist internally or externally. Learning all of this up front will allow you to figure out how much you have to do to prepare everyone for the journey and how easy or difficult that journey will be for them. You can do this by setting up a meeting or workshop with your stakeholders or by sending them a survey. I always prefer in-person meetings or calls, so that I can read everyone's body language and probe if I see something that seems interesting and that I might need to know.

Situational Awareness

To begin the diagnostic, we want to know what's happening today, where the stakeholders are going (or where they *want* to go), and where they have been. And we want to determine how to use that information to help clear the path safely for all of us to do the work. In aviation, we call this *situational awareness* (SA) (*https://oreil.ly/AA3M4*). When you are flying a plane, many factors are constantly moving and evolving as you move through the sky. There are weather systems and flight systems, and the failure of one of those systems could lead to disaster. In operational terms, SA means having an adequate understanding of the current and changing state of a system and being able to anticipate future system or environmental variations or developments. It's an essential element for safety, enabling pilots to *perceive* (understand their environment), *comprehend* (synthesize that information), and *anticipate* changes in the environment so that they can make informed decisions about the path their plane has to travel. Fundamentally, this is no different than what we do in Futures Thinking. But in this initial diagnostic phase, we are using these principles to determine the possible future of the project itself by anticipating potential issues we could face along the journey. Creating your initial diagnostic of a situation is really an attempt to gain a type of situational awareness so that you can properly prepare your material for the project. Some questions you may want to start with are:

- Why does your audience want to look into the future, and what are their goals?
- What kind of project will futures be applied to?
- How are they doing short- and long-term strategies today?
- Who is in charge of strategy today?
- What has or hasn't worked in the past?
- How much do they know about Futures Thinking?
- Who are your potential allies or skeptics?
- What or who are the potential barriers or threats to the success of the project?
- How open is the organization to change, innovation, or wild ideas?
- What are the vocabulary and artifacts they are using for strategy today?
- How do they think about the future today?

WHY DOES YOUR AUDIENCE WANT TO LOOK INTO THE FUTURE, AND WHAT ARE THEIR GOALS?

Getting to know the *why* will help you define what the deliverable might be. It will also give you some insight into what your stakeholders' goals are. What is driving their interest in Futures Thinking? Do they need a new vision? Are they compelled by some event or competition in the market? Are there new threats to the business? Do they see an emerging technology that they want to take advantage of? Or is an emerging technology becoming a threat? Do they want to use a new process to map out a long-term strategy? Try to pinpoint what the directive is. This will help you define the focal issue (the subject matter you are investigating), which we'll cover in Chapter 5.

WHAT KIND OF PROJECT WILL FUTURES BE APPLIED TO?

There are many potential applications of futures. Is this a business strategy project through which you'll help define a long-term strategic roadmap? Or is it an innovation project, in which you'll look into the future to imagine a next-generation product? Or are you using futures to understand potential future markets and customers? Each application will need a different mix of methods and deliverables. Determine the exact parameters as early as possible so that you can assess what will be necessary, including timeline, budget, resources, and methodology. Maybe you'll be teaching the members of a team how to use the toolkit so they can apply it to a range of projects. Find out what those projects will be so that you will know what to highlight and how to frame the process so it can directly apply to those projects later. Maybe the team members have done strategic projects in the past but without using futures. Or maybe they have projects that are not as focused on long-term outcomes, but they would like to have a process in place to pursue more strategic projects in the future. Whatever the case, learning where the process will be applied will allow you to tune the training or activities around what the final output might be. For example, your client might only want to learn about designing future products and is less interested in scenarios. While future scenarios are important to consider when designing for the future, you may want to spend more time training the team members on the design phase and then educate them later on the importance of scenario planning to inform future ideas.

HOW ARE THEY DOING SHORT- AND LONG-TERM STRATEGIES? AND WHO IS IN CHARGE OF THAT FUNCTION TODAY?

What does the word *strategy* mean to them today? Do they have strategists or analysts in-house already? This topic can be a little sensitive if there already are experts who are doing short- and long-term planning, but not in design or product groups; they could feel threatened. The truth is that futures work can look similar to traditional business strategy. We use different methods, but the output is similar. We also gather data and trends about today's markets and use that information to project a perspective about future conditions. If you can find out how your organization or client does similar work today, that will give you some insight into the similarities and differences, and how you might be able to fill in any gaps with your process. If you do run into a situation in which strategists are being territorial or skeptical, try to figure out why and turn them into allies rather than opponents. Show them the process and ask for their help and partnership. Be open to learning about their job, and hopefully they'll be just as open to collaborating with and learning from you. Then bring them onto your team and have them contribute as experts. Whatever the situation, try to minimize the risks of overlapping functions and unnecessary conflict wherever you can.

WHAT HAS OR HASN'T WORKED IN THE PAST?

Surely everyone has had to deal with the integration of new processes. Maybe it was a new onboarding process, or having to learn Agile or Scrum.[1] Even brand-new startups have had to wrangle their operation together using new tools or ways of working. Try to find out when a new process was last implemented. Did it succeed or fail? Was it introduced internally or by an external party (such as a consultancy)? Are there particular individuals or groups who can be your allies or barriers? Maybe the success of an implementation was due solely to a particular stakeholder who had the courage and passion to drive the initiative. This could mean that you need to ally with that person or get some notes from them about how they made it work. *Something* has worked in the past, whether it's a similar new process or another kind of implementation (of new policies, procedures, or software). Adapt that victory to facilitate yours.

---

1 Agile is a project management philosophy typically used in software development that allows teams to develop projects in small increments. Scrum, one of the many types of Agile methodology, is known for breaking projects down into sizable chunks called sprints. In software design organizations, design and engineering are usually trained in Agile and Scrum as a collaborative process to deliver software quickly.

HOW MUCH DO THEY KNOW ABOUT FUTURES THINKING?

Some people are more prepared than others for futures. Nobody needs to be an expert or a sci-fi enthusiast to be a futurist. But asking your audience what they know about the process or about similar futuring processes might give you an idea of how they will react to your approach. Are they excited about Futures Thinking but have never used it before? That means their hearts and minds are at least open. Or are they still a little cautious and skeptical? They might be inviting you in as an experiment to see whether futures work generates real value. These situations could require a lot of care and caution on your part. The "literacy" of your audience could be what makes working with them easy or difficult. In fact, there is a whole discipline and movement called *Futures Literacy* (FL).[2] Riel Miller, head of Futures Literacy at UNESCO, coined the term, defining FL as "the skill that allows people to better understand the role of the future in what they see and do. Being futures literate empowers the imagination [and] enhances our ability to prepare, recover and invent as changes occur" (*https:// oreil.ly/3DN15*). If your audience is futures literate, then how literate are they? If they have some idea of what you'll be doing, that will mean less preparation in the beginning and less hand-holding throughout. Futures Literacy (FL), as UNESCO describes it, is a holistic set of qualities that makes one capable of confidently grappling with the process and outcomes of futures work:

Innovation
FL makes it easier to innovate and to take advantage of innovations.

Discovery
FL makes it easier to detect and make sense of novelty, shocks, and surprises.

Leadership
FL diffuses initiative and experimentation throughout the community.

Strategy
FL makes it easier to see genuinely distinct strategic alternatives.

Agility
FL enhances the speed with which changes are perceived and choices are made.

---

2 In the next chapter I'll discuss the futures mindset, an attitude toward and basic understanding of and comfort with Futures Thinking that is useful in preparing people for futures work.

Confidence

FL makes changes easier because it makes change more comprehensible.

Knowing

FL embraces multiple ways of knowing the world around us, including emotions and contextual specifics.

Resilience

FL makes it easier to take diversification approaches to both risk and uncertainty.

Choice

FL makes it easier to build choice menus that are more diverse.

Capability

FL empowers exploration and invention that take advantage of uncertainty and complexity.

WHO ARE YOUR POTENTIAL ALLIES OR SKEPTICS?

Getting to know your stakeholders and identifying allies and skeptics is very important, because they are the operational crew on your vessel into the unknown. Be wary of people within the team or client side that are not fond of futures work or have opinions and agendas that are contradictory to yours. These could be people who are focused only on near-term goals, people who don't believe you can plan for the future because it's too unpredictable, or people who don't want to spend the budget on work that can't accurately be measured (this, by the way, is a myth; futures *can* be measured, and I'll address how in Chapter 11). Doing a quick stakeholder map or Power-Interest Matrix (Figure 3-1) and/or conducting some interviews could help you identify these key players. A *Power-Interest Matrix* identifies who in the organization has power (financial, influential, managerial) and their level of interest or investment in your efforts. Due to the potential sensitivity of this diagram (some may be triggered by the realization that they don't have much power), you should be careful about whom you do this exercise with or to whom you show the results. The diagram should primarily be for your eyes only as you navigate a potentially complex ecosystem of people who could affect your project.

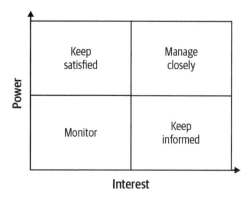

Figure 3-1. Power-Interest Matrix (source: adapted from an image by Nielsen Norman Group)

WHAT OR WHO ARE THE POTENTIAL BARRIERS OR THREATS TO THE SUCCESS OF THE PROJECT?

No project is without hurdles. There are always potential barriers you may need to breach or work around. It's just a matter of how high or how encompassing those barriers are. A barrier could be a skeptical engineer, a jaded vice president, a process that looks like yours that was a failure in the past, a lack of funding, competing timelines, competing agendas, or difficult attitudes in general. Take note of these during your diagnostic, and try to understand them as early as possible or set up ways to facilitate your path around them. These could include leveraging your allies and champions when a conflict arises. For example, if you are a consultant and you suddenly get skepticism from someone on the client side, you may not have authority to deal with the situation effectively. Thus, calling on a leader from the client side to deal with the situation and reignite their trust could be your lifeboat. Or there might be a lack of funding to execute the full process you really wanted to do. You may need to call on a leader to try to secure more funding. Try to be resourceful and always think of how you can generate the best results without selling yourself or the process short. It is possible. You may need to take shortcuts, but it doesn't mean failure is imminent.

HOW OPEN IS THE ORGANIZATION TO CHANGE, INNOVATION, OR WILD IDEAS?

The word *innovation* can mean different things to different people. Can you ask them for an example of something they've done in the past that was wildly innovative? What might seem innovative to you could be just a carbon copy of their competition. These are the boundaries you need to find early on so that you know what is safe to discuss or suggest and what is an unsafe or uncomfortable territory for them. Maybe they like being uncomfortable and disruptive. Maybe this is a team that thrives on taking risks. Is there a history of leaps and bounds in innovation somewhere, or has this organization been fairly conservative the whole time? Will they appreciate you pushing them outside their comfort zone, or will you have to play it safe the whole time? Were there certain people or groups involved in making that previous innovation a success? Or was it purely a top-down decision? Their answers will help you dial in how you might want to frame innovation down the road.

WHAT ARE THE VOCABULARY AND ARTIFACTS THEY ARE USING FOR STRATEGY TODAY?

Speak their language and get to know the artifacts they use for strategy. What are the words they are using, and what kinds of documents are shared to represent strategic work? Are they PowerPoint presentations with charts? Or rich data visualizations and narratives? Everyone is likely to have a strategy document somewhere. There might even be activities they are doing today that are similar to futures, but they don't necessarily call them that. Vocabulary will be extremely important throughout this journey. Unless they're already being used, you may not want to introduce the words *speculative* or *fiction* to avoid inaccurate associations too early (*Design Fiction* could seem like *science fiction*, thereby giving the perception that it is fake and for entertainment purposes only). Think about how they talk about strategy, innovation, planning, design, and their customers and research. Use those same terms to build their trust or until you feel it's safe to start introducing other terms.

Here are a few terms to listen for:

- Strategy
- Innovation
- Vision
- Long-term planning
- Long view
- Opportunity or threat assessments
- North Star
- Stretch goal
- Future of X
- Emerging (markets, customers, technologies)
- Blue sky
- Green field
- Skunk Works
- BHAG—Big Hairy Audacious Goal

Using their vocabulary not only will make futures work feel familiar and relatable to them but also will allow you to have a vehicle of communication that will make things easier to explain. You can use this vocabulary to make analogies to futures work, draw connections between their processes and yours, and then slowly start to introduce them to new terminology. There might be situations in which you never say the words *futures* or *foresight* and refer to it only as a "vision" and "strategy." If that is what they are comfortable with and what resonates with stakeholders, then go with it.

HOW DO THEY THINK ABOUT THE FUTURE TODAY?

Before introducing futures, I always try to ask people how they think about the future. What does the word *future* mean to them, and what inspires them today? Do they see the future as being 1 year, 5 years, or 50 years from now? Where do they look to learn about the future? Is the future positive (transformational) or negative (collapsing)? How *their* future is perceived can tell you a lot about their imagination, their aspirations, and their nightmares. For many, the future is a story about technology (flying cars, AI, robots); this is partially due to stories given to us within science fiction narratives. If you find that their views of the

future focus on tech, you may want to push them to think beyond that and help them discover other considerations, such as policy, culture, and environmental impact. If their future is colored by what the competition is doing (i.e., they describe the future based on competing products and don't seem to look beyond current market trends), then you'll know there's an opportunity to explore beyond their domain for inspiration. For many who are not practicing futures, they tend to be influenced by what they see on social media, in films, and/or on TV. That isn't necessarily wrong, but it can be constraining during the ideation phase. Opening up their minds to alternate views of the future can inspire their imagination and introduce concepts they may never have considered before.

Closer Than We Think

In the late 1950s, Arthur Radebaugh, an American futurist, illustrated a comic strip for the *Detroit Sunday Times* called *Closer Than We Think*. Every week Radebaugh depicted a vision of daily life enhanced by future technology (see Figures 3-2 and 3-3). Radebaugh wasn't necessarily a trained futurist, but he certainly was a professional illustrator and "imagineer" (*https://oreil.ly/n_OtZ*).

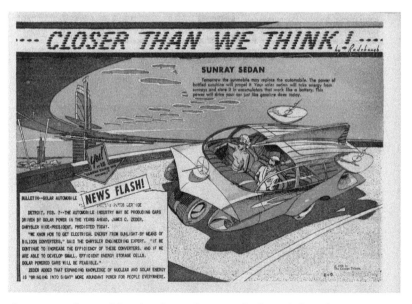

Figure 3-2. Arthur Radebaugh's Closer Than We Think *comic from the* Detroit Sunday Times, *February 9, 1958—"Sunray Sedan"*

Figure 3-3. Arthur Radebaugh's Closer Than We Think *comic from the* Detroit Sunday Times, *April 27, 1958—"Robot Warehouses"*

Radebaugh created hundreds of *Closer Than We Think* comic strips but never quite received the attention he deserved; he died quietly, uncelebrated, in a veterans hospital in 1974. Only recently has his work resurfaced, permitting us to discover where he was right and where he was just a little off. Imagining the future is never a precise science, but it gives us a chance to visualize something so that we can have a conversation about it. It didn't matter at the time whether Radebaugh was right or wrong; it was about dreaming of potential futures and opening up hearts and minds to the different possibilities. "We all dream of a better, brighter, more exciting future where the wonders of technology are there to serve and entertain us," says Todd Kimmell, the director of the Lost Highways Archives and Research Library, which is dedicated to American road culture, and Radebaugh "made that fabulous world of tomorrow seem practically at our fingertips."

Another illustrator who received acclaim for his visions of the future and of far-off worlds was Syd Mead, who was responsible for the concept art and visual language for science fiction films such as *2001: A Space*

Odyssey, *Star Wars*, *Tron*, and *Blade Runner*. Before becoming a sci-fi illustrator, Mead worked for companies like Ford Motor and United States Steel. In the 1960s he published a book of futuristic illustrations for U.S. Steel (see Figures 3-4 and 3-5) that was intended to promote steel use by automakers. He had creative freedom to depict anything he wanted, as long as each image showcased the use of steel. Much of his work seems to depict a very distant future and to traverse the boundaries between fantasy and science, and it has inspired generations of scientists, filmmakers, and futurists.

Figure 3-4. Syd Mead, U.S. Steel Series

Figure 3-5. Syd Mead, Palm Springs 2006

But the future doesn't always look like the depictions in movies or books. Think about what today looks like compared to 10 or 20 years ago—sure, we have electric cars, mobile phones, and drones, but we're also still using a lot of things that we had back then. Much of our world continues to operate on archaic technology (in some places, paper documentation is still the standard). Our world doesn't ever fully transform into a space-age epic; it's accretive and built upon lots of the mundane technology we've always used.

In an article on *Core77* (*https://oreil.ly/IIrWn*), Nick Foster, RDI, former head of design at X (formerly Google X), describes the "future mundane":

> Take a look around you, it's likely that you're interacting with a contemporary piece of technology, be that a smartphone, tablet or laptop, but take a look further around the room. There may be things which are older, things which come from another time—an LED TV atop a vintage table, a Playstation next to a '60s vase, an iPad in a leather bag. If industrial design is in the business of making stuff, then we need to understand that this stuff piles up, favela-like. Humans are covetous, sentimental and resourceful;

they cling to things. When we render the future as a unique visual singularity, we remove from it any contemporary hooks. When designing a new screwdriver, it's important to remember that it will probably sit in a toolbox filled with other tools, perhaps inherited from a previous generation.

Getting an idea of what these visions of the future look like to your client or audience will allow you to choose how to open their minds to different visions later. Do you want to show them something a little more provocative? Or do you want to show them something they are familiar with first and then expand from there to push their perspectives of what the future could also be? When people think about the future only in terms of sci-fi, it's an opportunity to show them that there is a practical and ethical future that relates to culture, policy, nature, and people.

Methods for Situational Awareness

When thinking about how you're going to design your diagnostic (your series of questions to gain awareness of the current situation), you can consider several approaches. A simple survey or conversation can be effective. But maybe you want to do it as a workshop to get multiple stakeholders in the room. Workshops are great for having a more communal conversation and exposing deeper hopes and fears. But be careful: while workshop environments are great for collaboration, they also can surface sensitive power structures that could limit what people are willing to say. Popular opinions could rule the room, or subordinates may not want to mention problems in front of managers (in some cultures this is very common). However, these events could be a chance for you to gain deeper insight into teams and to gain support for the transformation you hope to create.

Here are just a few examples of traditional methods you can use:

- Sailboat Exercise
- SWOT
- Premortem
- Business Model Canvas
- Causal Layered Analysis

SAILBOAT EXERCISE

I already introduced the idea of sailing to an uncharted island, so this exercise ties in neatly with that analogy. Originally designed by Luke Hohmann as a way to gather and organize customer complaints, the Sailboat Exercise has some very useful symbolism that can be employed as analogies for strengths, barriers, risks, and goals. The sailboat in general can represent the ship you want to sail into the future. The route you are sailing is the series of exercises you need to traverse to get to your destination. The route can also be your map or directional guidance system that points you toward the future.

To run the Sailboat Exercise, draw a sailboat (this is the company, team, or project) on a board or on a piece of paper (Figure 3-6). Next, draw anchors extending from the boat into the water (these are the issues that hold the ship back). Ahead of the sailboat are rocks (barriers or threats), and beyond the rocks is the island (their goals and visions). Last, draw the wind blowing the sails (the forces that are propelling the boat forward). Have your participants describe what the wind, anchors, rocks, and island are in their current world. You can use their comments to probe more deeply into the issues and identify opportunity areas for yourself or the project.

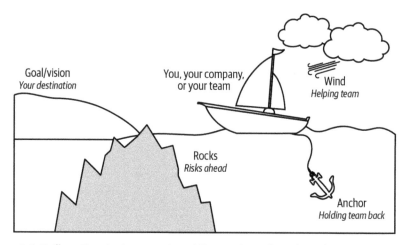

Figure 3-6. Sailboat Exercise (source: adapted from an image by Luke Hohmann)

The sailboat itself is just an analogy; you can replace it with any type of transportation vehicle. The boat could easily be a spaceship, and the goal an uncharted planet; the wind becomes the ship's thrusters, the anchor is the cargo

the ship has to drag, and the rocks could instead be asteroids or other obstacles ahead.

There are several benefits to using the Sailboat Exercise:

- The exercise is easy to explain.
- It offers a playful analogy that can ease tensions around potentially sensitive subject matter.
- It surfaces useful information around drivers and barriers in the organization.
- It highlights future aspirations and threats.

There are also some cautions associated with this method:

- The exercise may reveal only superficial issues; you would need to understand why certain views are occurring (e.g., some may believe funding is a rock, while others might not).
- It could be seen as too playful (the analogy may seem to childish to some).
- It can be subjective or based solely on who is in the room and might not consider deeper issues.

SWOT

SWOT is a common framework used for many types of assessments, such as corporate strategy, internal and external risk assessments, marketing and financial planning, product strategies, and team building. On a board or on paper, draw four boxes (Figure 3-7). The boxes represent the *strengths, weaknesses, opportunities,* and *threats* of the topic or objective you're interested in. SWOT is a very common framework in strategic planning, as it is easy to draw and explain and is also a good collaborative exercise. It's also a chance for people to complain or get excited about what's ahead. As with the Sailboat Exercise, you may need to read between the lines to understand how a strength or weakness could assist or impair your approach.

Let's say you're working with a bank that has never used futures before. The bank is strong in delivery but not so strong in innovation. It might be known to follow the competition and deliver competitive products, but rarely does it lead the market with fresh ideas. Its weaknesses might include a conservative CEO or team; they may get things done but aren't as creative as they want to be. Knowing this information about influential actors or teams who could

hinder your process would be useful when doing your planning. Conversely, understanding what strengths drive innovation or team excitement could be a model for how you want to package Futures Thinking for them. Not only can SWOT surface the mechanics of the business that are working or not working, but it also can identify allies or skeptics early. All this information will still need to be synthesized to allow you to prepare your approach (the methods you will use) and the deliverables (what makes sense for this team in its current state that will allow it to fully take advantage of futures in transforming its operation, products, or strategies).

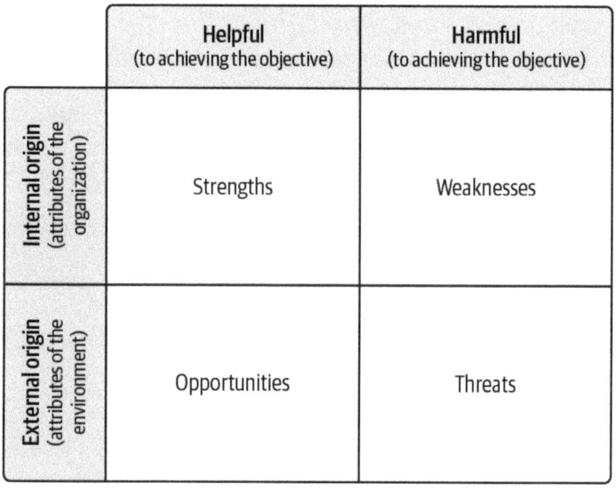

	Helpful (to achieving the objective)	Harmful (to achieving the objective)
Internal origin (attributes of the organization)	Strengths	Weaknesses
External origin (attributes of the environment)	Opportunities	Threats

Figure 3-7. A SWOT analysis identifying strengths, weaknesses, opportunities, and threats (source: adapted from an image by Xhienne on Wikipedia (https://oreil.ly/IazkR); used under CC BY-SA 2.5 (https://oreil.ly/ajAOL))

The SWOT framework offers several benefits:

- It is a fairly common framework in business.
- It's easy to draw and explain.
- It can expose key stakeholder allies or skeptics.

Some cautions around using the SWOT framework are:

- It may require deeper investigation into root causes or explanations.
- People might not understand how it relates to futures work.
- It can be subjective, depending on who is in the room.

PREMORTEM

A *premortem* is the inverse of a postmortem. Postmortems retroactively examine what may have gone wrong with a project and are used to discuss how you might improve next time. A premortem asks what *could* go wrong with this project before it even starts and how you might avoid getting into such situations before they happen. The exercise could begin with a simple inquiry of "What could go wrong?" Or you could ask: "What is the best-case scenario for the success of this project?" You can come up with as many categories of questions as you'd like. The idea is to document any hopes or fears about the project.

Let's say that the bank I mentioned earlier has a few stakeholders who don't fully support the futures effort. The project sounds exciting, but they don't believe the teams will use it; it could just die out without proper implementation (this happens a lot in consulting, by the way). You may want to use the premortem to understand how you can make sure the project succeeds (Figure 3-8). In this case, should you try to secure additional funding and resources for the implementation now? Or is this project dead in the water before you even attempt to pursue it, because the majority of perception is negative, and there is no time or money to support the full implementation of future initiatives down the road? The premortem becomes a futures method within itself to help you identify and mitigate those threats as a team.

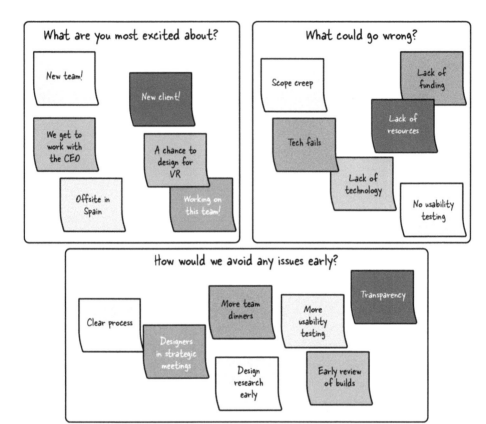

Figure 3-8. A premortem exercise (source: adapted from an image by Atlassian (https://oreil.ly/vPll8))

The benefits of using the premortem method include the following:

- It's easy to draw and explain.
- It's great for early risk assessments.
- It allows you to troubleshoot issues early and create contingency plans.
- It has flexible categories (you can have as few or as many questions as you want).

There are cautions associated with its use as well:

- You could risk introducing overly negative themes when discussing potential risks.

- You could spend too much time hypothesizing about failure rather than how to make the project succeed.

- It could seem oversimplified—sometimes surface issues mask deeper root causes for why certain things succeed or fail.

CANVASES

A canvas is merely a structured series of buckets (boxes) used to capture multiple types of information. The most common one is the *Business Model Canvas* (*https://oreil.ly/lEqtN*) proposed in 2005 by Alexander Osterwalder, who deconstructed business models into their key elements and their relationships (Figure 3-9). The nine building blocks are used to identify the pillars of success and how innovation might play a role in transformation. By identifying critical dependencies such as costs, revenue stream, key partners, and value propositions, you can discover what is truly important to a business's strategy succeeding now and in the future. Even if your work isn't specifically targeted toward transforming a long-term product or business model, the Business Model Canvas can be a starting point for understanding how a business is working. Thus, running this exercise as a *current state* analysis can be the primer for how you might want to set aspirations for a *future state*. In some cases, a client or team may not be able to answer all the questions, so make sure you have the appropriate people in the room. It's also important to disclose how you are using the canvas so that others know it's just for data gathering, as opposed to solutioning (some stakeholders could get an itch to solve problems as they state this information). If they do begin to ideate solutions, just ask them to hold back and wait until you complete the process so that you can have a much richer discussion considering the trends, scenarios, and other aspects about the future that could be important in creating transformational visions.

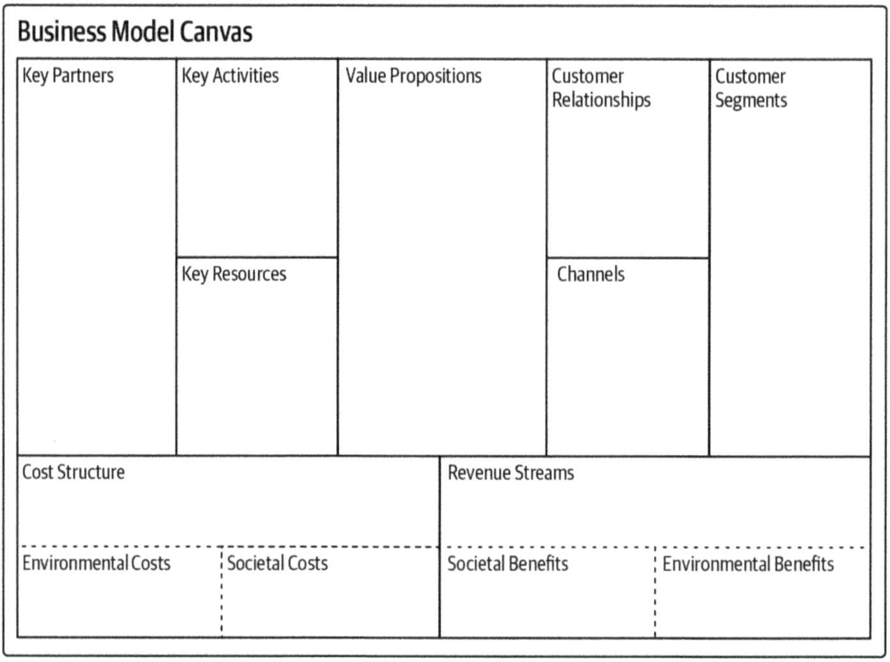

Business Model Canvas

Key Partners	Key Activities	Value Propositions	Customer Relationships	Customer Segments
	Key Resources		Channels	

Cost Structure		Revenue Streams	
Environmental Costs	Societal Costs	Societal Benefits	Environmental Benefits

Figure 3-9. Business Model Canvas

Variations on this idea include the *Lean Canvas* and the *Lean UX Canvas*. These are similarly meant to document what is essential for a system to operate without excess expenditures. They can include conditions such as users, user benefits, business outcomes, customer segments, and value propositions. Keeping this in mind as you walk through the futures process with stakeholders will allow you to help them focus on where they want to go and what might need to transform along the way.

Let's say your client or team wants to discover new markets or take advantage of emerging technologies like AI in the next five years. But its current business model shows no signs of competency in this field. The canvas can help expose these gaps in its current capabilities and help suggest investments in upskilling the organization so that it can be prepared to utilize AI down the road. While there is definitely a structured hierarchy present within some of these canvases (what are the baseline needs, and what supports those needs), you can also use canvases in an unstructured way as a data-gathering tool. You can create a canvas for anything, really. Since it's just a framework of interdependent boxes, you

could create a canvas populated with all the diagnostic questions I proposed earlier and use that in a workshop rather than a series of interviews.

Consider these benefits of using canvases:

- It's great for understanding deeper details of the business and where futures might be applied.
- It's also useful for exposing interdependencies, strengths, weaknesses, or opportunities within a system.
- It gives you the flexibility to create categories of your own.

And consider these cautions associated with this method:

- It could require more time to facilitate.
- The client might not want to expose sensitive business information.
- It could oversimplify complexities in a system or model.

CAUSAL LAYERED ANALYSIS

A common framework in Strategic Foresight is *Causal Layered Analysis* (CLA), which can be used anywhere within the foresight process. I am introducing it here because I believe it can be useful as a starting point for situational awareness. It can be used to identify problems and the potentially deep-rooted systemic pillars that influence or inform the collective view about the problem. Created as a futures research technique by Sohail Inayatullah in 1998, the tool has been widely used to investigate topics such as the global financial crisis, terrorism futures, global governance, aging and the changing workforce, climate change, and water futures in the Muslim world.

Sohail Inayatullah, UNESCO Chair in Futures Studies at the Sejahtera Centre for Sustainability and Humanity, explains the Causal Layered Analysis this way:

Causal Layered Analysis is like an archaeological dig. You peel back the layers of discourse and dialogue to get at the root assumptions and myths that underlie them. By uncovering these hidden layers, you gain a deeper understanding of the issues and open up new possibilities for action and change.

CLA intends to unravel how culture and systems shape the opinions we have today. It's very similar to another exercise called the *Five Whys* in which you ask the question "Why?" several times to uncover the root cause of a situation or problem. CLA categorizes the causes and effects into the layers of issues that give rise to the *litany*, or what we accept to be true. In CLA, there are four layers (Figure 3-10):

The litany

The tip of the iceberg, everything we see and accept to be true: the data, the headlines, the issues people talk about up front.

Causes

The immediate underlying reasons we believe make the litany true. These are the systems in place that reinforce the litany.

Worldview

The assumptions, perceptions, and personal worldviews.

Myths and metaphors

The narratives that support the views above. There may not be any real justification for myths; they just exist without validation.

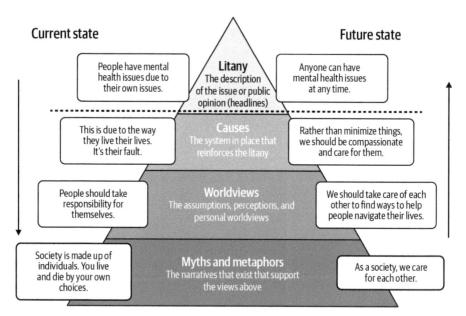

Figure 3-10. Causal Layered Analysis (source: adapted from an image by Scalable Analysis (https://oreil.ly/7G-8q))

To conduct a CLA exercise, you begin by moving down the left side of the iceberg, starting at the litany and moving down into the myths and metaphors. Transformation discussions then ensue as you move back up the right side of the iceberg from the myths and metaphors layer toward the litany. As you move up, you discuss how to change the narrative—how to debunk the myths, worldviews, and systematic layers that have created the current litany—until you eventually resurface with a transformed litany at the top. The CLA can be a lengthy exercise to run, so be sure that you have the time and the proper stakeholders in the room when using it, especially if you are using it early in the diagnostic phase.

Like any other suggestion, use CLA wisely and cautiously. Determine how you will facilitate this exercise. Will your audience be able to understand these concepts easily? Or might they get confused between what a "worldview" is versus a "myth or metaphor"? Do you need to eliminate a layer or change the definitions to make the exercise simpler? Beneath each layer are systems upon systems that support the layer above it. Whether you call them myths or worldviews, they are just deeper systems and deeper root causes. Even the iceberg analogy is just a symbol to illustrate the layers that are deeply hidden beyond your view. The reason this exercise can potentially take a long time is because people might not realize what the layers are beneath a particular belief or system. And as you get deeper into the exercise, they may not want to address the cultural myths that permeate and underpin their opinions without them even knowing it. In the end, the exercise can expose a lot of issues that drive people to think and behave a certain way and provide them with huge opportunities to transform through futures work.

The benefits of using Causal Layered Analysis include the following:

- It is useful as a root-cause analysis exercise.
- It supports the development of more powerful and richer future scenarios.
- It develops shared visions of a preferred organizational future.
- It can jump-start the foresight process.

And there are cautions too, of course:

- CLA requires more time and an understanding of the vocabulary.
- It requires honesty about problems, which can be limiting depending on who is in the room.

- It may constrain action through "analysis paralysis."
- It requires an experienced facilitator.

In Closing

The key to understanding your audience lies in your ability to be diligent about intelligence gathering as early as possible. If you don't have the budget for workshops or interviews, try to find a cheap or free and easy way to do it. Think of it this way: the people you are working with will be putting their trust (or at least their business) in your hands, so you probably want to get to know them a little and make sure they are comfortable with your navigation abilities and with being on your ship—and vice versa. We all want to get to that uncharted island successfully, so we need to make sure we are all on board. The questions you ask will help you discover what to do or what not to do and will pave the way for you to introduce the more complex principles later.

The Futures Mindset

Have you ever tried explaining what you do for a living to a complete stranger? No matter what your occupation, there's always a moment in which you have to pause to think about how you're going to explain your job. As a designer and consultant, I've often had to teach a client about Design Thinking or explain what kind of futurist I am. I might start by asking if they're familiar with the term *Design Thinking* and gauge their level of knowledge about structured, methodological inquiries that lead to the design of products and services. Then I might introduce the word *futurist*, but I would be very careful not to give the impression that I'm a science fiction writer. I might use the words *innovation* or *strategy*. Once I'm able to gauge how deep I can go with my explanation, I begin telling them more about what I do. In essence, I set the stage for the discussion; I prepare myself and them for a deeper conversation about what I do. There is a similar conversation that ensues when you begin futures work. In this chapter, I'll discuss some tactics for priming your audience for futures—how to discuss key principles of futuring, show them examples, and assess whether they are ready for the real work that is ahead. I call this adopting a futures mindset.

A Mindset and an Attitude

After you have completed your initial diagnostic and are able to confidently understand who you are working with and why, it's time for some additional preparation. Remember, we haven't started the actual futures work yet; we are still in the early stages of planning how we want to work with our client or team. I've also mentioned Futures Literacy and how it could enable people to grapple with future concepts and open their minds to embarking into ambiguous and uncertain environments. In this chapter I'll discuss a similar concept called the *futures mindset*—a state of mind and attitude that will allow your audience to prepare for the process and journey mentally and emotionally. Adopting and living with a futures mindset not only makes the job easier for everyone but also allows them to practice a sense of awareness of emerging patterns in the world at all times. Because it's not just about understanding the principles; it's about embodying an intuition and developing a muscle so that you can brave the world you want to explore. It allows you to scan for patterns around you, in front of you, and behind you, so that you can anticipate the future as it arrives. Without understanding and accepting some of these core principles, you run the risk of falling back into behaviors that can limit innovation and progress (the anchors holding the sailboat back). Ultimately, the futures mindset is a combination of awareness and attitude, and what you seek is to gain the trust of your audience by establishing the following:

A positive and optimistic attitude
> This seems obvious, but having a positive attitude about everything you are about to do, no matter how bleak or dystopic your perspectives or ideas might be, will allow you to forge through some of the most difficult parts of the process. As your data and research grow, the process can become very overwhelming, so keeping everyone's spirits up while you work through the analysis and prioritization will help you get to the end with a more collaborative and optimistic attitude.

A basic knowledge of core principles and stages of the process
> As part of the mindset preparation, decide which theories, principles, or methods you want to expose now rather than later. Too much information in the beginning could be debilitating and could paralyze your audience before you even begin the hard work. But explaining that there is research, scenario planning, design, and strategy is certainly a good way to frame the stages of where you will be going. You can decide whether to talk about the activities in great detail; some leaders or teams may want that level of detail, while others may want you to just shepherd them through the process.

An acknowledgment of fears and a readiness for potential failures
> Failure and fears are imminent. You will explore and postulate on many future possibilities that may scare your team. It's natural for people to think negatively and to also become paralyzed by and complacent about threats that they cannot control or that could seem very bleak. Help them

realize that the future is not always pretty; it takes no political stance and can breathe life or wreak havoc. To effectively assess all of this constructively, we need to be careful how we emotionally process this information. If we let our fears or fear of failure consume us or hinder us from thinking about the future, then this journey will be a lost cause before we even begin it.

A familiarity with different types of outputs

Futures can generate different types of outputs: trend cards, scenario stories, images, videos, speculative artifacts, and 10-year roadmaps. People who are not familiar with futures work might not understand what the end goal looks like and what their options are. Show them a range of examples, or an example of what they think they want, so they know what to expect and to get them excited.

Support and excitement to work with you and be led by you

As a leader and facilitator, you want people to be excited about the process and the outcomes, and about working with you as their spirit guide. Establish a positive rapport with people as early as possible to establish that trust and support system. This will improve collaboration, attentiveness, and advocacy when you need it.

Priming people to adopt a futures mindset can be tricky. We already discussed how using their vocabulary will help set the stage for what futures is (or isn't). There are a few concepts and frameworks that you can begin introducing. You can show all of them or select what you believe will resonate the most. Here are a few places you can start:

- Explain the characteristics of a futurist
- The future is always VUCA (volatile, uncertain, complex, ambiguous)
- The Futures Cone
- Analogies of a journey
- Recognize and embrace fears and failures
- The Futures Triangle
- Show relevant and appropriate examples

Explain the Characteristics of a Futurist

As the captain of the ship, you may want to begin by telling people a little more about yourself, your background, how long you've been practicing, or even some of your accomplishments. This will give some credibility to your skillset and allow you to share some stories about how you work, things that have happened in the past, and what to expect from your leadership along the way. If this is your first time using futures, you may not feel entirely comfortable calling yourself a futurist just yet, but don't worry—in learning and using the process, you are inherently becoming a futurist, as well as training others how to be one. Still, it's important to explain (and set some disclaimers about) what a modern-day futurist does. First and foremost, we are not wizards or oracles, and we don't have a time machine or a telephone to call the future. We use our tools to reveal possibilities and facilitate strategies to optimize for the futures we want or don't want.

Here are some additional qualities of futurists:

We don't predict.

We don't tell you the future; we offer possibilities. If you were in the business of *predictions*, you could end up backing yourself into a corner, where you'd have to defend yourself against things that you didn't get precisely right. Inherent in our genes is the ability to anticipate changes and prepare for them. The best we can really do is visualize different outcomes and, through our analysis, describe the probabilities of certain events. It's up to us to decide which ones we want to manifest or defend ourselves against. So instead of making a prediction, we use vocabulary that expresses the plurality of the future. We help people see the different alternatives so that we can make informed decisions about where we want to go (or don't want to go).

We focus on patterns.

Our work doesn't just come from having a wild imagination and watching and reading science fiction (though that is surely one type of inspiration). It comes from experience and from rigorous analysis of trends and other patterns we see emerging everywhere. We use research, quantitative and qualitative data, and anything else we can observe or get our hands on to see the similarities, differences, and forces that are changing the world.

We have an open mind and embrace uncertainty.

Being a futurist requires an ability to welcome the uncertainty and possibilities of the future. In the next section, we'll talk about how the future is always VUCA: volatile, uncertain, complex, ambiguous. Knowing and embracing VUCA allows us to look at the positive and negative sides of the future objectively, which in turn permits us to address anything with confidence, logic, and rigor.

We harness our imagination.

A good futurist draws on many sources of inspiration but also has the freedom and agency to harness their imagination by asking "What if?" questions. Imagination and creativity are key ingredients when trying to discover new ways to innovate in future worlds. Some organizations struggle to challenge the status quo and lack the courage to entertain new ideas. But only by nurturing the imagination can we really begin to move out of the shadow of doubt and competition and brave new worlds.

We look around us, backward and forward.

Futuring requires that you look backward into history as much as you look forward in time. If we forget the lessons of the past, then we are basically starting fresh every time, and that makes for pointless and useless work. There's as much to learn about where we've been as there is about where we are going. There's also a ton of data to mine from the past. Futurists aren't just forward thinkers; they are students of history as well.

We are collaborative.

"Thinking about the future is a collaborative and highly communal affair," says Marina Gorbis, executive director at Institute for the Future. "It requires a diversity of views. We need to involve experts from many different domains." We futurists seek out multiple sources and perspectives to gather our intelligence, and we never work in a vacuum. Without diverse views, our ability to envision the future we desire may be limited. However, when we incorporate different perspectives, we take into account a broader range of viewpoints, reducing the prevalence of self-centric assumptions. Since the future is inherently complex, we must work together to build the world we want.

We are inclusive.

Sometimes I reference the Overview Effect when discussing the responsibilities we have with futures work. The Overview Effect (*https://oreil.ly/m7t9L*) is a cognitive shift that affects some astronauts when they see Earth from space. Many say they no longer identify with a specific nationality or culture; instead, they see themselves and everyone on the planet as one people, living on one world. A good futurist honors that this world is made up of many living entities (humans, animals, and plants alike) and that our planet is our one home. Once you unlock the potential of Futures Thinking, it can inspire you to harness it for broader social and environmental impact. Unfortunately, there are some people or organizations that are not interested in inclusivity or sustainability. They might be interested only in revenue or in how to impact their bottom line. Futures Thinking is also inherently political and can be controlled by those who have the power, finances, and resources to create the futures that they want. It's a constant negotiation and at times a war zone. There are always debates and conflicts and competing interests when it comes to what the future should or shouldn't be. But don't let that discourage you from advocating for the values you want in a project. Our responsibility is to guide. If you see opportunities to expand views, take them; if not, it's not the end of the world (hopefully)—but continue to practice as a shepherd of the process. Much like in Design, we do our best to help our users by also doing the right thing for society and the environment; sometimes we get that chance, while at other times we solve for only a user-centric problem.

Describing our craft in terms that people understand while also accentuating some of the qualities that make us practitioners and experts in our field is just one way to introduce people to the futures mindset. We want to slowly introduce our audience to how we operate and open their eyes to the possibilities of the future. They don't necessarily have to learn and embody all these things immediately, but it can help set the stage for how we will operate. This will allow them to feel more comfortable with the process so that they won't be too surprised should any confusion occur along the way. That's not to say that there won't be any surprises—there always are! But at least they'll be more prepared to welcome them and can hopefully use them to fuel the process rather than hinder it. Can you imagine being on a boat with someone who is terrified and skeptical of your ability to sail? They might eventually become a nuisance and could potentially spread fear to others. This is why it's important for everyone to understand that it is a safe space to explore—that there's a reward at the end and you can help them get there, but it takes a dedicated crew.

The Future Is Always VUCA

One of the first acronyms I introduce when priming for a futures mindset is *VUCA*. VUCA stands for *volatile, uncertain, complex,* and *ambiguous.* The term was first introduced in US Army War College documents as far back as 1987 and was subsequently cited in several other War College documents relating to strategic visioning and strategic leadership. Futures practitioners use this term often when referring to the environment of the future. An ex-Navy colleague explained to me once that in a conflict environment, "Every step you take is a new VUCA environment, and you constantly need to reassess your situation." It's true. With each passing moment, new factors are emerging that could change the tides. A global pandemic or earthquake could occur at any time. Life in general can be complex and ambiguous. Similarly, in a business environment, a competitor could invent the next iPhone, or some TikTok influencer could go viral promoting a random product and disrupt an entire industry. That's business! It's volatile and uncertain. We don't necessarily need to think about the future as a constant war zone, but for some it is just that, and for some it's a matter of life or death for their business. The principle of VUCA can be used to acclimate people to the uncertainties of tomorrow. You can think of those uncertainties as dangers or as opportunities. But once you realize they exist and you can willingly and objectively embrace VUCA and tackle it with excitement and ease, then you are more prepared for anything that happens.

To recap the four elements of VUCA:

Volatility
> Unexpected or unstable outcomes exist and may be of unknown duration, but they are not necessarily hard to understand; some knowledge might be available.

Uncertainty
> Despite a lack of other information, the event's basic cause and effect are unknown. Change is possible but not a given.

Complexity
> The situation has many interconnected parts and variables. Some information is available or can be predicted, but the volume or nature of it can be overwhelming to process.

Ambiguity

Causal relationships are unclear. No precedents exist—you face unknown unknowns.

The Futures Cone

One of the first diagrams I show to people who are new to futures is the *Futures Cone* (Figure 4-1). Developed by Joseph Voros in 2003, this cone of possibilities describes time as an ever-expanding cone that can include alternate futures or realities. Many in the foresight and futures communities regularly use it to map out alternate scenarios that can play out over time. Often people tend to think of the future as a straight line; this is the default version of the future, which can look much like today, or "business as usual." Voros calls this the *Projected* future. But the great silent joke within our community is that *business as usual* means that the future not only looks like today but also inherits the same problems, with no surprises or change. We all know the future isn't always a carbon copy of today. It's always different (and also looks different from different perspectives). Change is inevitable, be it slow or hidden on the periphery.

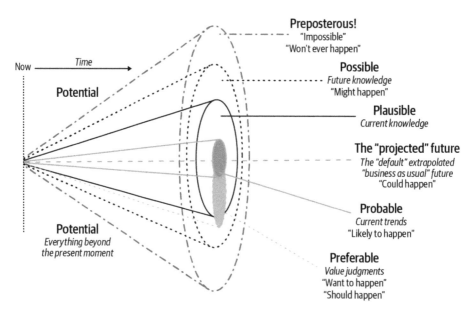

Figure 4-1. The Futures Cone by Joseph Voros (source: adapted from an image in The Voroscope *(https://oreil.ly/smt_S))*

I sometimes like to explain it with a flashlight analogy. Imagine you are shining a flashlight into a dark room (Figure 4-2). You have no idea what is in there; it could be threatening or completely safe. The future is similar. We have no idea what the future holds at first, but we try to see into it using our tools. When we shine a flashlight into a dark room, it illuminates some areas of the room so that we can see what's ahead. The brighter and wider the beam of the light, the more we see, and the safer we feel if no threats are visible. And if we *do* see something threatening, we can prepare. You can also use the analogy of traveling down a dark road or trail—but whether it's a dark room or a dark road, we all would love to have a flashlight to see, wouldn't we? However, there are always boundaries to the light. The beam's reach stops at some point, and we can't see anything in the area beyond it unless we move the light into that area, but there's still always a dark edge somewhere.

The Futures Cone similarly categorizes these areas, with the brightest area (the center) as the *Probable* or *Projected* future, and moving outward to the *Plausible* (could happen), the *Possible* (might happen), and the *Preposterous* or impossible futures. Futures that lie within the Preposterous cone are ideas that just can't exist because of some limitations—I sometimes use human teleportation or interstellar travel as an example of the preposterous (in our current year of 2024). These are fantastic ideas, but they are not achievable at present, and maybe not even within the next decade, with our current state of technology and understanding. But who knows—perhaps some new discovery or event could move one or the other idea into the Possible realm; until then, it lives in the darkness outside your cone of possibilities.

The cone isn't necessarily a method (though it could be used as one); rather, it's a principle that describes the multiple alternate possibilities that the future could bring. I introduce this idea very early when explaining futures to show that the future is not linear and that we have agency to imagine all types of scenarios and worlds. The future can be plural, wild, and exciting and is not prescribed by any one idea, roadmap, or destiny. Later we'll see how we can lay the Futures Cone over the backcasting method to determine how different variations of a vision could play out over time. Once your audience can accept that the future truly is a wide-open canvas that presents endless possibilities, they will have begun their first steps toward adopting a futures mindset.

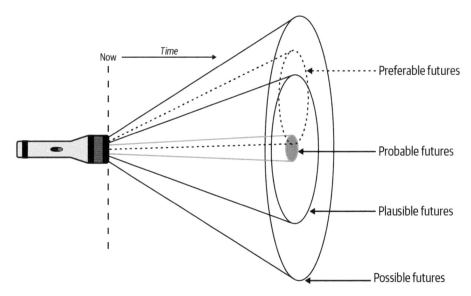

Figure 4-2. Futurology: shining a bright, broad beam of light into the darkness (source: adapted from an image in The Guardian *(https://oreil.ly/8RHvG))*

Analogies of a Journey

I've already introduced the analogy of sailing a boat to an uncharted island. I also sometimes use the analogy of planning to climb Mt. Kilimanjaro. For people who have families, you could refer to planning a trip with your children to a theme park. In that last example, you definitely want to prepare for every potential circumstance or event that could put your child's life (as well as your sanity) at risk. Depending on how you look at them, every one of these destinations would require a lot of planning—and in the case of Mt. Kilimanjaro, a bit of physical and mental training would be required as well. The same goes for futuring. No matter how much of the futures toolkit you plan to use, you will want to prepare everyone for the journey, and you'll want to know as much about your destination as you can, given the resources you have at your disposal. We want people to feel comfortable and prepared for the activities, and we want to visualize the best picture of the destination (the future world) so that we can decide how we want to operate and survive there.

The great thing about an analogy is that it abstracts the process into something that people can easily relate to. Try to find an analogy that makes sense for your audience. Maybe you want to use a story that relates to their current product cycle? If you are working with a software company and have discovered during your diagnostic that the people at the company are not very open to change or find it hard to relate to certain abstract analogies, maybe you'll want to talk about the process in terms of a one-year product roadmap. It is something they are likely very familiar with and will allow you to use the same language as them to get them on board faster.

Whatever analogy you choose, try to make it easy to explain so that you can link key facets of the process to ideas they know. Don't worry about holes in the story; just focus on the ideas. Sure, Mt. Kilimanjaro is not completely uncharted, and people have been there before, but there are still uncertainties around the weather, the wildlife, and even your own resilience on the hike. But that's not the point. The point is that it's a foreign territory full of surprises—much like the future!

Recognize and Embrace Fears and Failures

Failure in life is imminent. It will happen, but we must sail on nonetheless, or else we sink. In many organizations, fear manifests itself in many ways, such as fear of failure, fear of competition, fear of the CEO, or fear of losing funding, just to name a few varieties. All are valid fears, but they can become obstacles to transformation and innovation. Try to understand where the fear lies during your diagnostic and find a way to quell that fear immediately, if possible. Controlling and managing fears is always a little difficult to plan for. Futures is not a precise science; this is why we try not to *predict*. Yes, we can be wrong! Think about how you might introduce the idea that failure is acceptable. Should you discuss how the process could fail (perhaps during a premortem)?

We don't necessarily set out to fail, but we should be open to the fact that we can and probably will fail at some things or be inaccurate somewhere along the way. And that is OK! In fact, design is rooted in failure. Our imperative is to build, test, fail, learn, and build again. Everything we have and own today is the result of improved failures. So yes, this process, as rigorous as it can be, can result in failures. This is a tough concept to sell at times. Some organizations don't like to be too negative or pessimistic because they don't want to expose uncomfortable uncertainties. Some consulting firms get paid millions of dollars to deliver the "correct" answer. And they sure do work hard to sound confident. Confidence can sell anything (if you're good at it). But the reality is that you sometimes must balance between optimism and *cautious* optimism (if you want to frame it that way). It makes sense that too much pessimism can spread fear and doubt. But you can't simply ignore potential threats to a system.

One way is to reframe them with a different word like "challenges" or even invert them as "opportunities to address gaps." If we looked only at the positive opportunities and never addressed the threats, then we would be blindsided every day and would continue to waste time and money recovering from threats we never planned for. We don't necessarily like surprises either. We want to be as prepared as possible, and we want to do anything we can to control our world. So, by thinking about failures and fears with a problem-solving mindset, we can be more effective when we do fail. If we establish that we equally welcome opportunities and failures, we can decide how to deal with them more effectively by creating safety nets so that we don't fall as hard when we do fail. A safety net could be a conversation about what we will do if something isn't correct or fails, or it might be a contingency plan with some invested resources, procedures, and policy to recover from a failure, should one happen. Again, we do this every day when we plan out our daily lives. Some fears we can repress without deterring our entire plan, and others we have to really think strategically about and figure out how to deal with them in a constructive and optimistic way.

The Futures Triangle

Another common framework used in futures work is the *Futures Triangle*, which was developed by Sohail Inayatullah in 2008. It describes three forces that are always acting on us at any point in time (Figure 4-3). The three points of the triangle represent the weight of the past, the pull of the future, and the push of the present.

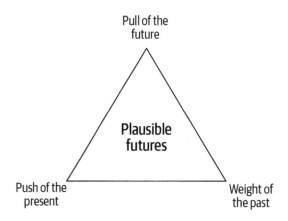

Figure 4-3. The Futures Triangle

In an article on the KnowledgeWorks website (*https://oreil.ly/bzdZz*), Maria Crabtree, director of Strategic Foresight projects for KnowledgeWorks, promotes the use of the Futures Triangle to "organize the possibilities ahead" without "succumb[ing] to the urgency of addressing current events," explaining that the Futures Triangle

> *helps us grapple with the convergence of drivers of change anchoring the past, developing the present and shaping the future. These drivers of change include forces in addition to...major societal changes [happening today].... The juxtaposition of different combinations of these drivers of change results in multiple futures for us to consider and prepare for.*

THE WEIGHT OF THE PAST

All the things that hold us back (the anchors of the boat) can be considered the weight of the past. Those who refuse or are reluctant to try new ideas or explore new ways of doing things might be encumbered by legacy systems, mandates, or their own strict viewpoints—these also represent the weight of the past. These views anchor people to a history that they don't necessarily want to change. They might be used to doing business as usual and are OK with that (also known as the "if it's not broken, don't fix it" attitude). This attitude isn't necessarily wrong, but if you are trying to innovate or transform your operation, strategy, or products, something will have to give. Being stuck too far at this end of the triangle or being too influenced or driven by the past will create an uphill battle when you are trying to introduce new processes or ideas.

THE PULL OF THE FUTURE

The pull of the future can be represented as the vision of the island you want to sail to. The pull is your aspirations, dreams, goals, or desires; it's the ideas that exist in the future that make you want to chase after them. There is always a pull, a place that you want to go to but may not know how to get to yet. These forces make us want to move forward and seek change or transformation. One of the first steps for any organization that is taking on futures work is to recognize that there is a future it wants and that it wants help getting there. Recognizing that is one thing, but wanting to do something about it is another. Many organizations might see that there is a future that is threatening or opportunistic but don't believe enough in a process to strategize around it. Those organizations may just want to "wait and see." In such cases, you may just need to wait and see also—you'll have to wait for them to call you when they are ready.

THE PUSH OF THE PRESENT

The pull of the future and the push of the present may seem similar conceptually. To determine the push, you'll have to distinguish between what drives you forward today and what is more of a vision or dream in the future. Some forces are the gas in the engine moving the business down the road. What trends, policies, technologies, or resources do you see and feel around you that are propelling your business or life forward? For some it is a will to keep the lights on (i.e., keep the business afloat). For others it's an obligation or expectation set by someone else. We do our job because it pays the bills and someone tells us to do it. It's part of the machine that we are sitting in that is naturally moving through time, as opposed to the pull of the future, which is something we conceive or dream that we want—a technology, a new policy, a more lucrative product, an easier job.

You may have noticed that there are similar concepts embedded in many of these examples. The Sailboat Exercise is similar to the Futures Triangle, which has some similar theories to Bill Sharpe's Three Horizons (which is discussed in Chapter 7)—all are intended to detail forces of change or existing barriers and complexities. How you illustrate these concepts for others will determine what people comprehend and accept. Any of these can be used as a way to explain Futures Thinking or as a discussion tool during the diagnostic. Consider how each of these examples is framed, and think about what's necessary to show and how to explain it. It can be a slow discovery for some, but eventually people will understand and walk the path with you.

Show Relevant and Appropriate Examples

As you become more invested in futures work, it's a good idea to keep a library of futures examples: speculative vision videos, articles, reports, roadmaps, process artifacts—anything and everything to represent futures that might come in handy one day. In my discussion of the diagnostic phase in Chapter 3, I talked about understanding how people think about the future today. Do they describe the future in terms of technology, culture, or the environment? Everyone's future can look and feel different, so it's necessary to get a baseline of how your audience sees the future today so that you can decide what's appropriate to show them from your example library. Some examples might be too sci-fi or too conceptual. Others might look and feel exactly like their products today or focus

on the same issues they focus on (such as AI-driven trends). If you show them something that is overly academic or conceptual, you could lose them quickly. But if you start with something they know and can relate to, you can then slowly expose them to other kinds of futures work. If it's a software company, maybe you show them future visions of software first. Then show them a physical product to explain different formats of futures design. Eventually they will be able to read between the lines and see how it's not necessarily the thing of the future but the *process* for how that thing was imagined that counts. And an investment and belief in the process is where you want to go by showing them different examples that resonate with them so that they can imagine what their end product could look like. You'll know that they get it when you see their excitement in what you show them.

In 2015, Microsoft's futures division released a six-minute video called *Productivity Future Vision* (Figure 4-4) that shows a future of Microsoft touchscreens, wearables, and other speculative technologies. It walks us through different environments, from a classroom and a corporate boardroom to a retail store and a science lab, and highlights how various transparent and interactive interfaces will play an important role in our work and lifestyles. This can be an effective video to show people, even though there definitely is technology that is on the edge of the cone of possibility, such as 3D holographic displays (Figure 4-5). It serves as an aspirational vision of where Microsoft as a company could be headed with its software, hardware, and operating systems. People also will recognize the Microsoft brand, so it can immediately add a level of credibility to the story.

Figure 4-4. An image from Microsoft's Productivity Future Vision *(https://oreil.ly/tdeA_): an executive views data on a transparent touchscreen wall*

Figure 4-5. An image from Microsoft's Productivity Future Vision: *a researcher analyzes data on a 3D holographic display*

Let's say you're working with a startup software company that builds an app for booking flights. The company may already have a vision and roadmap for the next two years, and the app has an impressive set of features that have been synthesized from numerous user interviews. But the company knows it wants to be more innovative and become the top booking platform in the next three to five years. Its current vision might be entirely influenced by its current roadmap, and there's nothing wrong with that. But if you want to use Futures Thinking to open up more possibilities, you might need to show the company what an alternate future could look like. You could play it safe and show other examples of similar airline booking platforms with different features that this company's app doesn't have. But that would just be a competitive analysis. Or you could show the company something a little more provocative, something that challenges its ideas of what a platform could do, and other things it could focus on that address emerging technologies, shifting trends, demographics and cultures, climate change—ideas that push the company outside of a commerce platform and make it think about social/cultural opportunities, new economic models, and challenging current policy frameworks. With that you are likely to find more disruptive opportunities. First ask yourself: is this an audience that would benefit and become inspired by a different kind of example? Or do you need to slowly wean them onto more provocative ideas? If you have primed them well and have opened their minds to uncertainties and new possibilities, then maybe you can start to push their imagination outside of their comfort or knowledge zone. Futures projects can be tricky that way. Some people really want to be provocative and innovative but in the end are simply too risk averse (this may be due to their organizational, cultural, or team environment, or a result of top-down directives and constraints).

The examples you use can really shape the way a company thinks about the future and what is possible with Futures Thinking. But you have to choose carefully. That's why the diagnostic is so important. Knowing how far you can push your audience and how willing they are to entertain provocative notions can allow you to sail them into exciting (and sometimes uncomfortable) territory. The reward could be lucrative. And taking these kinds of calculated risks and succeeding is where innovation can really thrive.

Another example I like to use is a vision video created in 2018 by GE Transportation, a GE business unit responsible for manufacturing locomotives and freight trains (Figure 4-6). An innovation team led by Ryan Leveille, then Head of Experience Design and Digital Innovation, used Futures Thinking and emerging technologies like virtual reality to help the company imagine a 10-year roadmap for its freight division. Among the deliverables was an animated video about how GE could leverage predictive analytics, drones, and other software tools to optimize its operation and respond to emerging threats that many freight companies were facing at the time amid the rise of online shopping and shipping. The technology presented in the video was not necessarily revolutionary for that time, but its application in an industrial setting was. The concepts the video presented were logical and accessible and didn't feel too far out of reach for the stakeholders. These characteristics are important for demystifying futures and for showing how practical futures can be realized. It isn't all science fiction.

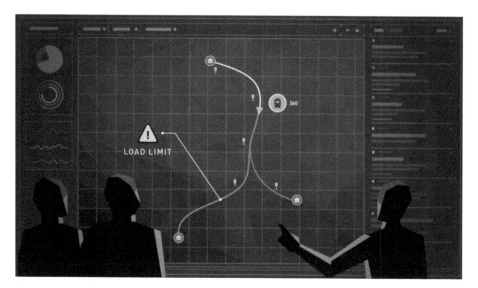

Figure 4-6. A still from GE Transportation's Future of Freight *vision video (https://oreil.ly/z0JFp)*

Some futures work can be pretty visceral and may not represent a future someone wants, but it might represent an important conversation that needs to happen when discussing future trends and implications. A more provocative (and sometimes uncomfortable) example that I sometimes show is by Agi Haines, from her project *Transfigurations* at the Royal College of Art in London. I've used this example for years to represent a practice called Critical Design (also referred to as Speculative and Critical Design), which I'll discuss in more detail in Chapter 9. For her exhibit she fabricated several *designer* babies, each one with modifications tailored for different purposes. Figure 4-7 shows a baby with extra folds on its head, which will allow it to work longer hours in hotter temperatures. Haines was fascinated with the manipulation of the human body and did extensive research into how we have been trying to change our physical form to adapt or adhere to cultural traditions over decades (Figure 4-8). In her artist's statement for *Transfigurations* (*https://oreil.ly/kGKq_*), Haines observed that "[the] human body consists of practicable elements that can be easily manipulated and engineered. Through surgical procedures our bodies can be stretched, shifted and sewn, yet still be functional. What then would stop us searching for a higher level of function than we have now? Especially if it may have the potential to benefit the younger, more vulnerable and more malleable generation?"

Modern plastic surgery provides us the ability to modify our skin, augment limbs, and even reassign our genders. So the technology exists, but its application to children is not ethically accepted (at least not today). While these babies are not real, they raise important questions about what we could do with technology if we had the abilities and desires. Is this the future we want? Or do we need to ask ourselves how to steer humanity away from this path? The baby, as a discourse, becomes a signal and an artifact from a potential future that provokes the viewer to investigate the implications of how design could play a role in our children's lives.

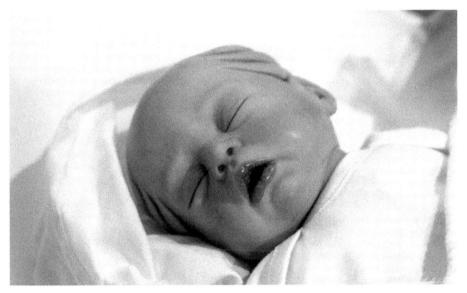

Figure 4-7. Agi Haines, Thermal Epidermiplasty, *from her exhibit* Transfigurations: *"'Transfigurations' depicts designs for potential body enhancements that have been surgically implemented. Each modification is put in place to imagine how these techniques could 'solve' a potential future problem for the baby, ranging from medical to environmental to social mobility issues, but at what physical, mental, social and economic cost?"*

Figure 4-8. Agi Haines, Podiaectomy, *from her exhibit* Transfigurations: *"A high incidence of asthma can be prevented by the removal of the central phalange. Leaving the soft fleshy skin exposed for the potential contraction of a hookworm, a parasite known for reducing allergic responses."*

Throughout the book, I'll reference more examples that could be useful in different situations. Some of these will be final designs, while others might be process outputs. Maybe you want to show process artifacts such as a scenario card or trend card, a backcasting map, or a Futures Wheel (we'll discuss these in detail in the following chapters) to describe how exercises are structured and how they can eventually lead to a more polished vision. Be careful, however, about showing low-fidelity artifacts too early. While they may be legitimate output from workshops, artifacts that are sketches or unrefined drawings of ideas could give people the wrong idea of what "final" output looks like. On the other hand, showing them something of low fidelity could lower their anxiety about having to be perfect or polished within their ideation or co-creation activities. Treat these situations on a case-by-case basis. Showing something that doesn't look polished could create skepticism, thus countering your efforts to impress and gain buy-in and trust. Not to say that a low-fidelity output is bad, but make sure your audience understands what it means before you show it. In the end, it will be partially up to your gut instinct as to whether an example is too risky

Modern plastic surgery provides us the ability to modify our skin, augment limbs, and even reassign our genders. So the technology exists, but its application to children is not ethically accepted (at least not today). While these babies are not real, they raise important questions about what we could do with technology if we had the abilities and desires. Is this the future we want? Or do we need to ask ourselves how to steer humanity away from this path? The baby, as a discourse, becomes a signal and an artifact from a potential future that provokes the viewer to investigate the implications of how design could play a role in our children's lives.

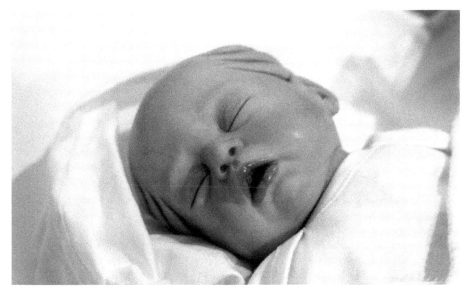

Figure 4-7. Agi Haines, Thermal Epidermiplasty, *from her exhibit* Transfigurations: *"'Transfigurations' depicts designs for potential body enhancements that have been surgically implemented. Each modification is put in place to imagine how these techniques could 'solve' a potential future problem for the baby, ranging from medical to environmental to social mobility issues, but at what physical, mental, social and economic cost?"*

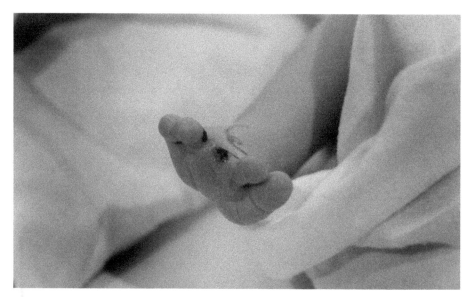

Figure 4-8. Agi Haines, Podiaectomy, *from her exhibit* Transfigurations: *"A high incidence of asthma can be prevented by the removal of the central phalange. Leaving the soft fleshy skin exposed for the potential contraction of a hookworm, a parasite known for reducing allergic responses."*

Throughout the book, I'll reference more examples that could be useful in different situations. Some of these will be final designs, while others might be process outputs. Maybe you want to show process artifacts such as a scenario card or trend card, a backcasting map, or a Futures Wheel (we'll discuss these in detail in the following chapters) to describe how exercises are structured and how they can eventually lead to a more polished vision. Be careful, however, about showing low-fidelity artifacts too early. While they may be legitimate output from workshops, artifacts that are sketches or unrefined drawings of ideas could give people the wrong idea of what "final" output looks like. On the other hand, showing them something of low fidelity could lower their anxiety about having to be perfect or polished within their ideation or co-creation activities. Treat these situations on a case-by-case basis. Showing something that doesn't look polished could create skepticism, thus countering your efforts to impress and gain buy-in and trust. Not to say that a low-fidelity output is bad, but make sure your audience understands what it means before you show it. In the end, it will be partially up to your gut instinct as to whether an example is too risky or is just the right one to get their attention or teach them something. Your diagnostic and conversations should tell you what to do. And you can always just ask someone on the team if they think an example is appropriate to show or if it's too conceptual or confusing.

Regardless of which models, examples, or diagrams you use to prime your audience for futures work, you ultimately want them to be curious and excited to embark on this journey. You might even want to do a quick pulse check periodically to make sure they are still with you and still on board with the process. Is there any confusion or skepticism about the principles, methods, or outcomes? Do you have new allies and champions? Even if you can get people only half the way there (eager enough to step into the boat with you and set sail), you're in a much better place than if you had no support or belief at all. Belief systems are important. We use them every day; we take the knowledge and memories we collect and place our trust in them so that we can make educated decisions about a system. But if that trust is broken, a bad memory is created. You could lose the team that you need to complete the journey. And sometimes it's hard to recover that trust based on the event or experience the team has with you. Use these examples to ignite that curiosity and desire for futures, and if you still aren't getting people on board, ask them what more they need to build that trust.

In Closing

These are just a few examples that you can use to help your audience understand the possibilities that Futures Thinking can generate and to help them adopt a futures mindset. Try to use your diagnostic research to tailor your conversation to their culture, vocabulary, and needs, and use this stage of priming to inspire and stimulate excitement for the journey ahead. In essence, what you want to accomplish is:

- Getting them more familiar with the process and concept of futuring
- Helping them understand the multiple forces acting upon us
- Introducing them to the different lenses we use to look at the future
- Preparing them for potentially scary, complex, and uncertain environments
- Motivating them to get excited about futuring with you

Remember, it takes experience, trust, faith, and a willingness to collaborate to see the future. You may not convince everyone, but that's OK; you just need to convince enough people to move the project forward and reduce the number of headaches and the skepticism (and potential mutiny) you might encounter along the way. This doesn't mean you'll have smooth sailing the whole way. You may have to rekindle excitement again at some point. And that just comes with the responsibility of caring for your crew as you sail into the unknown.

5

Framing the Problem

Now that everyone has adopted a futures mindset, you're ready to start gathering the project details for what you're going to be working on. The initial steps were for you to prepare yourself and whomever you're working with for this process. Now we need to define the WHAT of the project. Of course, as you run the diagnostic, you may certainly discover these details along the way. And they might have even been included in the initial project brief that was given to you when the client hired you or asked for your help. But now that we know more about the people, the context and application of the process, and the potential complexities and opportunities, we can start dialing in the specific details based on that information.

Having a clearly defined problem frame will serve you well because it will help you decide how to begin the initial research and start planning activities. In this chapter we look at how to articulate the question you're trying to answer. Futuring is more than an image we're trying to paint; it's the result of deep exploration into possible worlds. Similarly, we are painting that image using the tools at our disposal, as well as using our imagination to depict a world based on the limited information we have.

If someone said to you, "Paint me a city," you might paint any city that you imagine in your head. But that might not be the city they had in mind. However, with deeper inquiry, or given a more articulated proposition such as "Paint me a city in Spain in 2050 surrounded by sustainable buildings and flying cars," you might paint a much different picture, one with more specific details. The detail of it being a city in Spain tells you something about the geographic and cultural references you might have to include. Since flying cars are also in the statement, you might want to think about what the technology and transportation infrastructure might look like (especially if flying cars don't exist yet). Some of your painting will be based on memory, and some of it on pure speculation and imagination. Your *focus* becomes a little more precise due to the inclusion of specific details in the proposal.

Focusing Your Research

A *problem frame* is a set of parameters about the future you are trying to explore, and it can be as detailed as you need it to be to have some clarity about where to start and where you want to go. It's very similar to a project brief or project overview that's given to you at the start of a project. Typically, someone has an idea or a directive for a project they want to pursue and writes everything they know about that directive. The problem frame may be highly speculative or vague and not fully fleshed out, but it will at least give you a starting point for your research, and it may prompt additional questions that you could ask in the hope of refining it so that you can start.

In the diagnostic and situational awareness phase, you gathered information about the team or organization you're going to work with, so you know a little more about them—their sensitivities, goals, culture, and operation. Then you had to prepare them for the futures work by explaining what you'll do, why, and what the process and deliverables could look like. Now you are getting down to the real nitty-gritty details of the project. You'll frame the problem with the specific directive of where you are going and what to begin researching.

In this chapter we'll discuss:

- Determining the specific parameters of the problem area and your initial research goals
- Articulating a clear focal issue/question you want to solve
- Determining the time horizon
- Determining initial geographic boundaries
- Identifying the people (in the future), initial demographics or populations, users, customers, stakeholders, and so on

Framing the problem will require getting these project parameters really focused. It will mean synthesizing everything you've heard so far about where your team or client wants to go, where they've been, and how they want to get there. Knowing a little bit about their business will reveal how you might want to start this process and how to advocate for the outcomes you want to achieve.

Since part of the first inquiry (the diagnostic) is about the current state and how they work today, you begin this project-specific inquiry about the project by asking some of the following questions about the future they want to explore:

- What subject are they interested in exploring?
- What are the goals of the project?
- How far into the future do they want to look?
- What geographic region(s) are they interested in?
- Who are the people they are interested in (customers, users, their employees)?
- What market or industries are they interested in?

You may certainly get this information early. Maybe it comes as part of their request to have you work with them. But then you'll want to begin the diagnostic to understand more about their business, projects, and goals. I've placed this chapter here, as a third step of Stage 0, not because it has to happen after understanding your audience and futures mindset priming, but because even if you have an initial description of what your client or team wants to do, you will still need to do a little digging around before you lock this information down.

There is a scenario in which you decide to do a diagnostic, and what you learn completely changes your problem framing. For example, an automotive company might say, "We want to look at the future of mobility in the next 10 years. Can you help us use Futures Thinking to do that?" That is a broad statement, and you may have never worked with this team, organization, or sector before. So you naturally will want to respond, "Sure! But before we begin talking about the project, may I ask you a few questions about what you are doing today, how you operate, and how this project will ultimately benefit your goals and expectations?" After your investigation, you might find that you need to teach them a little more about the process to make sure it's clear they know what they're getting into. Once they are really excited and on board, you may need to refine the problem statement by adding more details:

We want to look at the future of mobility in San Francisco in 2040 for Generation Alpha. Specifically, we'll focus on micromobility on land and in the air (shared vehicles and other emerging transport) and the economic, cultural, and infrastructural challenges and opportunities in 2040. In the end we want a roadmap to tell us how to develop a vision for the next 15 years that allows us to be more sustainable using clean energy, growing our market to younger generations and expanding into new markets around the Bay Area.

This is a fairly detailed problem statement, with specific goals, demographics, and aspirations. It can help you dial in the first phase of research. As the project progresses and you gather more info about adjacent topics, sectors, geographies, technologies, and other trends, it can shift into something completely different. But what you arrive at after the deeper analysis will be much more refined than "the future of mobility in the next 10 years."

Alex Fergnani, a foresight practitioner and scholar, breaks down project framing into three components: Domain, Assessment, and Logistics (Figure 5-1). In his Domain Description, he includes the focal issue, the geographical boundaries, the time horizon, and the stakeholders involved. Some like to combine time horizon, geographic boundaries, and stakeholders or demographics into the focal issue statement. The problem frame should represent everything you intend to explore about the future you want to explore. Depending on your audience, you can configure it however you want. You could list the items separately from each other or write them together as a narrative or statement:

Domain Description

This involves deciding what kind of future scenarios are going to be created. For this reason, Domain Description can alternatively be called the *scope of the futures*. This component makes up the lion's share of framing. It is in turn made up of four elements: focal issue, geographical boundaries, time horizon, and stakeholders involved.

Assessment

Assessment involves deciding the purpose of the scenarios, i.e., what the scenarios are going to be used for. Assessment is different from Domain Description—the former has to do with the objectives of the organization in using those scenarios, while the latter relates to the content of the scenarios—although the two depend on each other.

Logistics

Logistics involves deciding all the technical particulars of the project itself. This includes decisions about the members of the scenario team, the duration of the project, the timeline and number of foresight workshops (if

any), the amount of resources that the organization is going to put into the project, and, most importantly, the futures and foresight methods that are going to be used, a decision that is dependent on the expected outcomes previously decided on.

Figure 5-1. Alex Fergnani's framing of a futures and foresight project (source: "A Detailed Account on How to Frame a Futures and Foresight Project" (https://oreil.ly/zRHke))

In Fergnani's framing, Domain Description, Assessment, and Logistics are all combined as one project brief or problem frame. For the purposes of this book, I'll break down the components of Alex's framework into two main sections: *problem framing* (this chapter) and *project framing* (Chapter 6). Problem framing will cover the Domain Description, and project framing will cover the Assessment and Logistics. This is just an alternate way of categorizing the info and activities, but ultimately with the same intent as Fergnani's framework.

Here I'll discuss just a few of the components that make up a baseline problem frame:

- Focal issue
- Time horizon
- Geography/region
- Stakeholders
- Demographics/population segment

THE FOCAL ISSUE

The *focal issue* (sometimes called the *focal topic, focal question, focal statement,* or *problem statement*) is the subject matter you are investigating in the future. It determines *what, where,* and *when* you are looking at and can be articulated as a simple statement or question. I think we all know what could happen if you set sail for a distant island without a GPS. The more precise your GPS, the safer you will feel about getting to that island. If your focal issue is too broad, you are just sailing in a general direction, and you'll be lucky if you hit the island. For example, if you are looking at "The Future of Work," that could mean a lot of things, depending on where you are located and what you want to focus on within the future of work (remote work, technology, human resources, cultural trends). You might end up spending a lot of time looking at many different aspects, making it more difficult to focus on a particular area that you might want to innovate in. But again, it depends on what the goals are. Maybe you do want a broad perspective at first. The focal issue can be a short or long statement. Some examples of what that could look like are:

- The future of work in Barcelona, Spain, in 2040
- The future of remote work in the banking industry in San Francisco, California, in 2030 for Gen Z

The focal statement sets the stage and the boundary conditions so that you'll have an easier time describing where you are going and what you are working on. Some focal issues are short and sweet and could even be used as the title of the project or report. I'll break down these components, but keep in mind that your focal issue can include other aspects that may not be mentioned here, such as domain or task specifics (e.g., the future of buying a flying car online for travel businesses in the Midwest region of the United States in 2030). And remember that a focal statement is a useful way of articulating the scope of your project. You could easily just list these parameters in a project brief. What's important is to identify them, document them, and use them appropriately to guide your project.

TIME HORIZON

The *time horizon* is the frame of time you are looking at in the future. The time horizon is important for several reasons: it gives you an anchor point in time to build a world around, it sets constraints and boundaries, and it gives you a horizon that can be agreed upon by all parties. Without a time horizon, you have no sense of where you're exploring in the future. It's like saying, "We're going to Europe for the holidays"—*where* in Europe are you going, and *when?* Because that information will influence how you plan, the clothes you bring, and the activities you schedule.

When determining time horizons, make sure everyone is aligned on the pros and cons of *near* versus *distant* horizons. I'll discuss this more later, but the further out your horizon is, the blurrier that world becomes. One thing I've learned is that if the horizon is too far in the future, the quality of the results may become too fantastic, lofty, or inaccessible. The exercise then produces results that may not seem tangible, or they may seem unmanageable or too difficult to design cohesive strategies around. You might say, we want to achieve carbon neutrality by 2050. Sure, that is a valid time horizon; a lot can change in 25–30 years, and it could take a country that long to reach that goal. But you may also want to set some near-term time horizons so that you can plan actionable concrete strategies to reach that future horizon successfully.[1]

Setting a far-term time horizon isn't wrong or dangerous per se; in fact, we want people to stretch their imaginations and generate grand visions and goals without constraints. But eventually we might end up scaling it back to something closer, something that is feasibly attainable in a shorter amount of time. The reality is that we must dream big but plan small (at first). Use your diagnostic to test the boundaries of what your audience is willing to dream of and how far they're comfortably willing to explore. Maybe your team wants to look at the next 100 years. If we develop a 100-year vision (much like the postcards of Jean-Marc Côté that we looked at in Chapter 1), that vision is surely aspirational and can be very fantastic. But it may not be particularly useful to today's generation other than as inspiration or a dream. When you're working with an organization for

---

1 Setting near-term and mid-term goals that align toward a long-term time horizon will be covered in Chapter 11.

which your goal is to develop a long-term actionable vision, 100 years might be too distant for anyone to connect to emotionally or operationally. As long as you are clear on *when* in the future you are looking, you are OK. The closer our time horizon is (1, 5, 10 years), the more accuracy we can attain through the data we gather, and the more realistic it will feel to reach those goals when it comes to designing the strategic roadmap. Personally, when starting with a new client, I never go beyond a 10-year time horizon. Ten years is just far enough out to push people to imagine new possibilities without today's constraints, but it's close enough that people can imagine what that change might look like. Sometimes I ask people to think about what the world was like 10 years ago and how different it was then. Now invert that and think about how dramatically different the next 10 years could be. A 5-year time horizon is even safer to work with because it is far enough into the future and yet close enough that it's much easier to see the strategic plan and rely on certain trends to still be there. Of course, many things could happen that would make either of these scenarios unreliable or unpredictable, but some stakeholders feel just a little more comfortable working with a nearer-term horizon as opposed to a 50-year horizon.

Thomas Chermack, a professor at Colorado State University and a scenario planning expert, uses scenario planning with a time horizon of no further than five years in the future. It really depends on the technology and the pace of change within an industry. Some industries move really fast, others much more slowly. The American writer Stewart Brand, cofounder and editor of the *Whole Earth Catalog* and founder of the WELL, the Global Business Network, and the Long Now Foundation, discussed his concept of "Pace Layers" in the 1999 book *The Clock of the Long Now* (Basic Books). His diagram of Pace Layers (shown in Figure 5-2) had this caption: "The order of civilization. The fast layers innovate; the slow layers stabilize. The whole combines learning with continuity." The six Pace Layer levels, in descending order from the highest and fastest to the lowest and slowest, are Fashion, Commerce, Infrastructure, Governance, Culture, and Nature. Stewart's intention was to give insight into how a healthy society works. Understanding the Pace Layers will allow you to provide some initial lens as to the rate of change that is typical within your industry or within the structures that influence your industry.

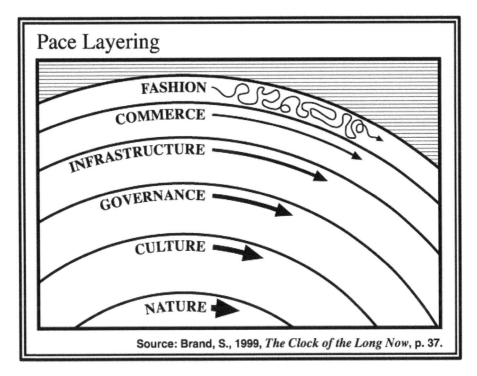

Figure 5-2. Pace Layers diagram from Stewart Brand's book The Clock of the Long Now

In a conversation I had with Alex Fergnani, foresight practitioner and academic, he talked about how industries can be impacted by major changes in a sector:

We always have to play with the contingencies of the organization. Sometimes you have an event coming up that will completely change the whole structure of the sector, the whole structure of the industry, possibly very soon. In some industries, you have regulation coming that is going to be disruptive sooner or later, it's going to come in the next year or so. Then you can apply foresight to that, which is a mix of contingency planning and scenario planning. If this regulation comes, then we use this scenario and do this. If this regulation doesn't come, then we'll use that scenario and do that. So everything is up to the context and the contingencies. There's no gospel. Of course, that being said, working with foresight over long-term horizons is usually better, especially for larger organizations, and especially if you're working in public service or in government.

Imagine that you are trying to see something in the distance as you survey your surroundings in a national park. Human vision has biological limitations. Even people who have 20/20 vision can see clearly only to a certain point, and looking beyond that point can become difficult (without binoculars). We can discern some things (shapes, patterns), and due to our memories we make relationships to what they might be (a mountain, a river, people), but until we move physically closer, we are relying mostly on our experiential memories and on some guesswork, as our mind tries to make sense of what we are looking at. Similarly, with foresight, we can try to discern ideas of what the future looks like. The closer it is, the more we can understand what it is going to be or look like (the Probable part of the Futures Cone), but the further out it is, the blurrier it gets. We can discern some idea of how history may play out, but the guesswork increases the further out we try to look. In my opinion, to make futures work, we have to find the sweet spot between a future we can believe in and touch, one that pushes our imagination and comfort, and a future in which we can have some anchor in reality so that we can use what we know to test the waters of future risks and challenges *today*. Again, we are not making predictions and trying to be correct, but we do want some security so that we can confidently set forth and make a plan we can all agree is worth following.

There are futurists out there who will probably disagree with me on what kind of time horizon is "right" for futures work; they may say that if the time horizon is less than five years, then it's not really futures work. I disagree. In fact, everything beyond this very minute is the future. And the premise of this book is learning how to make futures work in many contexts and situations. If you adhere to the idea that futures work operates only in a distant realm (10 years and beyond), you could have a difficult time gaining buy-in and advocacy among people who are new and averse to long-term thinking. This may be especially true with companies that are very focused on surviving the current economic climate or don't know how to understand the return on investment (ROI) of long-term thinking. It doesn't mean they don't deserve our tools. We have to reframe how futures can be useful for short-term strategies also. Ultimately, we want everyone to be thinking about the future for the sake of our businesses, society, species, and planet. The myth that has continued to propagate around who can practice futures and where/when/how they can do so has only contributed to the greater problem of successfully implementing futures throughout society, as many futurists say they want to do.

GEOGRAPHY/REGION

The future will always be different in different parts of the world, due to many factors. The way that people live and work in San Francisco, California, can be very different from how people work in Delhi, India, or in Barcelona, Spain. Each region is different due to social and cultural values, technology availability, economics, policies, and many other dependencies that drive how people go about their daily lives. So when you try to think of the future of transportation in Delhi versus San Francisco in 2040, those futures might not look exactly the same. This is not to say that a futures project can't involve several regions. It really depends on what you or your client wants to look at. You could look at the US versus China, but in the US, trends play out differently in different cities around the country. Be mindful about *where* you are looking and how that affects the work you are embarking on and the output you want. Maybe your client really wants the mega trend worldview. Organizations with branches around the world will surely want to look at global trends, but they should also look at local trends to figure out what can drive their individual local businesses. When we talk about trends in Chapter 7, we'll zoom into mega, macro, and micro trends and fads and examine how they differ at different geographic scales.

STAKEHOLDERS

Stakeholders are the various parties internal and external to the organization who are expected to have an impact on the futures examined in the focal issue. Stakeholders include much larger communities and key players in systems. This concept allows us to look beyond a particular user group and consider wider implications of both the world we are building and the multiplicity of actors and functions within it. Key stakeholders, of course, are those within the business who drive strategy, leadership, and funding, and all those who might be involved in implementing a strategy: engineers, designers, field workers, and their managers. But there might be other people or functions in the system external to the business. These could be policymakers, communities, competing businesses, adjacent industries, and their constituents. Articulating a few of these stakeholders in our problem framing statement gives us some initial direction as far as where to research and whom to talk to. For example, if we believe that policymakers in healthcare are critical to our implementing a suite of new services in 2030 or sooner, we may want to include them in our list of stakeholders and begin researching the policies that are holding back our vision today and might need to change in the future.

DEMOGRAPHICS/POPULATION SEGMENT

Let's not confuse demographics with traditional current state personas. In traditional design thinking, we conduct user interviews to gather information and then collate and synthesize that information to create a customer persona—a generalized profile of a particular type of customer or market.

When starting futures work, you may want to look at a particular segment of the population in the future. For instance, what is the aging population between 60 and 70 years old going to need in the next 20 years? Or how will teenagers play video games in the next 5 years? These communities can surely be identified at the beginning as part of the problem framing and can direct your research, but they can certainly be shaped and influenced by the scenarios of the future that you derive from foresight activities (which I'll discuss in Chapter 8). A teen in the midst of a pandemic in 2040 will behave very differently from a teen in a nonpandemic scenario.

When defining your problem frame, you may need some specification as to the humans or other actors you want to investigate in those future worlds or markets. Since most of our inquiry is a projection of what we know today, you'll have to be cautious with how you treat this part of your research. You may be so used to thinking about human-centered design thinking, which puts people in the center of your research, that you may find it difficult to think about context before people in futures work. Futures Thinking will require a shift to thinking about future environments and conditions as a priority. This doesn't mean that humans are not important; it's just a different approach to how we imagine the future. But that doesn't mean you can't incorporate a human perspective into your focal issue in the beginning. Anything that helps you focus your research is useful.

If today is 2024, then 2030 is six years from now. If we are interested in 30-year-olds in 2030, then those are actually the 24-year-olds of today. We could try to understand that age demographic now and think about how they are growing up, what technology they have access to, what cultural themes and social movements are driving their decisions, and how they behave with regard to the economy, politics, environment, and so on. This is the current state of 24-year-olds and gives us only one picture of today. A lot can happen in six years. We can also look at 30- to 35-year-olds today and understand how much people are spending on business-class tickets right now, and for what purposes. What matters most to them? What technology or policies or cultural drivers are driving their decisions? From these perspectives, we can slowly start to piece together a picture of how we think a group of 30- to 35-year-olds might behave in 2030. But again, how they behave will be dependent on the influencing climate of the future. That is why it's important to do the futures work to carve out the environmental picture that can help describe what our target group has at their disposal and what factors might drive their decisions.

Another approach, one that I'll discuss in more detail in Chapter 8, is to develop *future personas,* which are extracted *after* you develop scenarios; in other words, you would create your future worlds and scenarios *before* you articulate what the people are like, the reason for this being that people behave very differently in different worlds. For example, a person in the pandemic world of 2020 behaved very differently from how they might have behaved if we had not had a pandemic. Economic, cultural, and technological conditions can drive massive changes in how people think, do, and feel in that world. So even if we had an initial demographic or population segment we were interested in as part of our focal issue, we create a more specific persona profile *after* we develop the scenarios they might live in. For those of you who might be used to starting the design process with user interviews and personas, you may need to momentarily deprogram yourself as we delve deeper into the futures thinking process. By not putting humans at the center of everything, we also have a chance to think more about external factors, nonhuman actors (such as the environment), and the invisible conditions that typically are overlooked when we think about what affects humanity and vice versa.

Domain Maps

Earlier we discussed the components of a problem frame. If you are struggling to figure out your framing, an alternate method you can use is called a *domain map* (Figure 5-3). You may use this when you are searching for a focus or when you want to synthesize information from your diagnostic to help you map out areas of interest. Domain maps allow you to visually draw out connections among topics of interest. This can be done individually or as a group. Much like a mind map, a diagram in which information is represented visually, usually with a central idea placed in the middle and associated ideas connected around it like a spider web, each topic in your world of exploration can be connected categorically to show interconnected relationships. This will allow you to expose all the topics your client or audience might be interested in so that they can have meaningful discussions and discovery around which topics they would like to prioritize and investigate. If you are working with information from your diagnostic, you can pre-populate the diagram with what you've heard about where they want to go and their interests, deliverables, emotions, and current threats and opportunities.

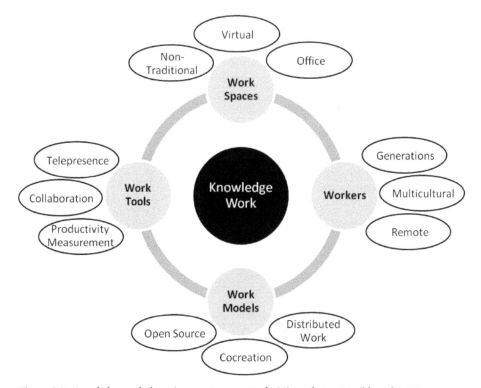

Figure 5-3. Knowledge work domain map (source: Andy Hines (https://oreil.ly/KdFz1))

Andy Hines, associate professor and program coordinator for the University of Houston Foresight program, advocates for the use of domain maps in every foresight project (*https://oreil.ly/KdFz1*), as it "helps build common understanding within the team and with clients, and also provides a launching point for research and scanning." After completing a domain map, you can work with your stakeholders to decide which are the most important areas of interest for the problem framing. You can do this by voting, or tagging topics by feasibility, interest, threat, or opportunity. Domain maps are essentially just a visual way to help the team focus on a particular area or areas to explore.

Problem Framing Canvases

As I mentioned in Chapter 3, canvases are a great way to capture information. As an alternative method, you can create your own canvas to capture the details of your problem frame. Each box can consist of the key components of the problem frame, and you can include additional questions, topics, or parameters that might help you focus your research. I've included short- and long-term goals in the example in Figure 5-4, as well as boxes to capture what people want to explore in the future, challenges, and enablers, to help identify other areas of interest that could inform the focal issue. These are just more ways to get people to focus on the future they want; you could add additional questions and use the canvas in a workshop or during your diagnostic as part of your initial inquiries.

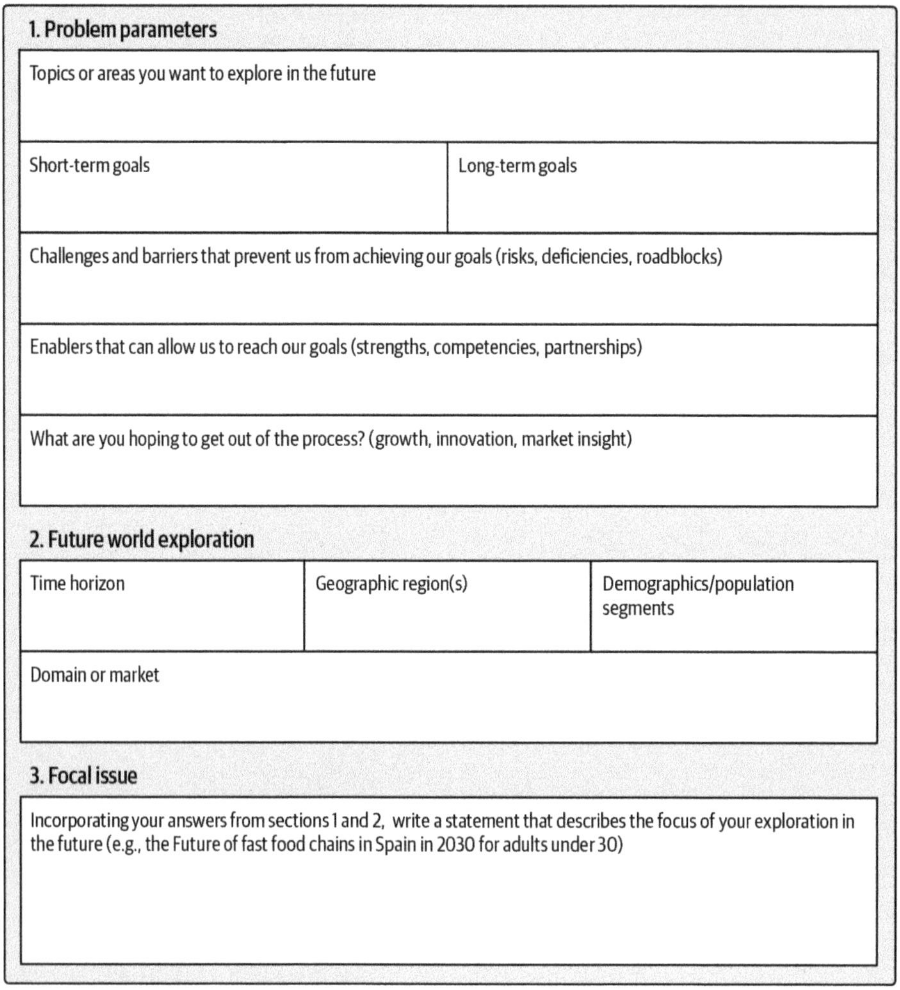

1. Problem parameters

Topics or areas you want to explore in the future

Short-term goals	Long-term goals

Challenges and barriers that prevent us from achieving our goals (risks, deficiencies, roadblocks)

Enablers that can allow us to reach our goals (strengths, competencies, partnerships)

What are you hoping to get out of the process? (growth, innovation, market insight)

2. Future world exploration

Time horizon	Geographic region(s)	Demographics/population segments

Domain or market

3. Focal issue

Incorporating your answers from sections 1 and 2, write a statement that describes the focus of your exploration in the future (e.g., the Future of fast food chains in Spain in 2030 for adults under 30)

Figure 5-4. Problem framing canvas

In Closing

Framing the problem not only gives you navigational direction for research but also focuses all your efforts downstream. When you have limited time, you will want as much focus as possible. You'll be surprised by how much easier things are if you know where you are going rather than just setting sail into the sunset and hoping to find an island. We can think of the ocean as being many potential areas to research before we set sail. The more we know about what's ahead and how to handle it, the safer we will feel, and the smoother our trip to the island will be.

6

Framing the Project

When it comes to setting up the logistics and parameters for a futures project, many organizations will have their own way of organizing the timeline, activities, and team. In this chapter, I'll mention a few items that you might want to think about or at least be aware of, as they will inform your activities and provide a safety net as the work builds up over time during and after the project. As both a facilitator and a futures practitioner (and captain of the boat), you will want to make sure you are directing, synthesizing, and delivering on the assets that the process generates. If you can afford to bring in others to assist, make sure they are fully aware of the process, define goals, and set expectations so that you can maintain guidance of the ship as you sail it through potentially turbulent waters. Design teams can have different configurations at different companies, so whether you are a lead designer, a director, or a strategist, the importance of keeping organized won't really be different than it would be with any other type of design-led engagement.

Project Logistics

If you don't have a product or project manager, you might be the one in charge, and that can work for or against you. Being the project lead as well as liaison, subject matter expert (SME), and facilitator is an excellent role for you as a futurist and leader. It gives you the opportunity to exercise management qualities and allows you to maintain control of every aspect of the process. However, managing your time and priorities could wear you out quickly. That's why having a project manager or assistant to handle some tasks and logistics can be valuable.

Here are a few activities to secure and understand:

- Map out the project timeline and budget
- Plan your key activities (research, methods, workshops)
- Secure your allies and champions
- Align on the deliverables
- Set up workspaces, repositories, and communication preferences
- Determine process success metrics

MAP OUT THE PROJECT TIMELINE AND BUDGET

Knowing your timeline and how much funding you have will give you a good idea of how to scope your work, methods, and outputs. These factors will vary depending on what you are trying to achieve and the deliverables. A trend report about the future of gaming in the UK in 2030 can cost less than a project that has six months to craft a future vision for healthcare that includes speculative products, services, and a strategic roadmap to develop those services. Timeline and budget can also influence which frameworks you'll have time to use and how much time you can spend in each stage. A good facilitator will use the timeline to orchestrate activities so they can seamlessly flow from one exercise into the next. If you plan too many activities, you risk burnout and could compromise the quality of your outputs. If you plan too little, you may not get the fidelity of output you are really looking for. Think about the amount of research, synthesis, or prototyping you might need to do. Will you have time to do it all, or will you have to make compromises somewhere? If you have to leave anything out, make sure it's clear to your team why certain decisions you are making are necessary for the timeline.

PLAN YOUR KEY ACTIVITIES

Once you know your general timeline (your route) and deliverables (your island), you'll have some idea of what you can accomplish. You don't necessarily have to use every method in the toolbox to get the job done. In fact, a good designer and futurist will cherry-pick the methods that are most appropriate given the focal issue, timeline, stakeholders, and deliverables they have. Being very concise about which activities you plan and how one input leads to an output is essential in making this an effective experience to remember. Think about what you want to accomplish and what is necessary. If you have time for only two activities, which would those be? How much do you need to do on your own as opposed to involving your client or team in a workshop environment? Maybe you can do the trend research yourself and use the client time to do prioritization and ideation workshops. Depending on what you are trying to deliver, research is a necessary and critical component since it is the input to your future scenarios and visions, so make sure you have ample time to include it. Then you can gauge how you synthesize that research into future perspectives by choosing some of the available methods and outputs, which I'll discuss in the following chapters. It's possible to sprint through activities, but like any design sprint, you are challenged by time and depth of research and quality of output.

It's usually helpful to start with the end in mind. What are the deliverables, and how can I walk backward from that point? What is the optimal approach to make sure all your deliverables are thorough and impactful? Ask yourself:

- How much trend research do I need or have time for?
- How much time do I need for each phase?
- Which methods will work best with this team and still provide the outcomes we need?
- How long will the strategic roadmap planning take?
- Are there complexities or barriers that could make planning and facilitation difficult?

Returning to our sailboat analogy: if we know that the island (the goal) is four days away but we have only three days of fuel, we need to make some adjustments so we get there faster and more efficiently—a combination of sailing with the wind and using a gas-powered motor. The same negotiations have to be made to make it to your goal without compromising quality for speed. Some methods could be rushed and others not. For example, if you rush your research and as a result are too broad and don't consider local macro and micro trends (which we'll discuss in Chapter 7), you might risk omitting key influencing factors in your future scenarios. If you rush your ideation, you might generate bad ideas, or ideas that look like things we already have today and aren't innovative.

It's also important as a responsible facilitator to make sure you are incorporating all the necessary meetings and alignment time so that everyone is on the same page. Make time for team huddles, breakout groups, share-outs, and final report-outs. With teams that are new to futures, I like to do frequent report-outs and check-ins to make sure that everyone is engaged in and understands the process and that they are acquiring new knowledge and ideas along the way. A simple pulse check at the beginning or end of key exercises can give you a sense of what is working and what needs to change moving forward to make the activities more effective. You might even consider doing such checks as casual conversations over lunch or dinner—or if you are remote, you could schedule a casual 15- to 30-minute check-in call. Remember that this is your vessel and that you want people to feel comfortable along the way so they don't mutiny or derail your plan. In Figure 6-1 I've outlined a general futures sprint of activities that could be distributed over a five-day timeline.[1] Of course, the context and scope of every project is different, and you definitely wouldn't want to rush your team through the process, but if five days is all the time you have, this is just one proposal for how you can plan your activities. This assumes that you've already gathered all your information about the team, client, or organization you are working with and that you have focal issue details already defined. If not, you might want to dedicate some time in your schedule for that phase of work.

---

1 A sprint (https://oreil.ly/L3mjf) as it is referred to in the text is inspired by the traditional Design Sprint (https://oreil.ly/IfiCG) invented by Jake Knapp when he worked at Google Ventures and is a five-day series of workshops inspired by Agile and Design Thinking, with the aim of reducing the risk when bringing a new product, service, or feature to the market. In this context, the futures activities are distributed across five days to show how you could potentially accomplish a rapid journey through the process. While this example might seem too quick to have deep conversations, it's meant to provide a template for how and when you might structure activities with limited time.

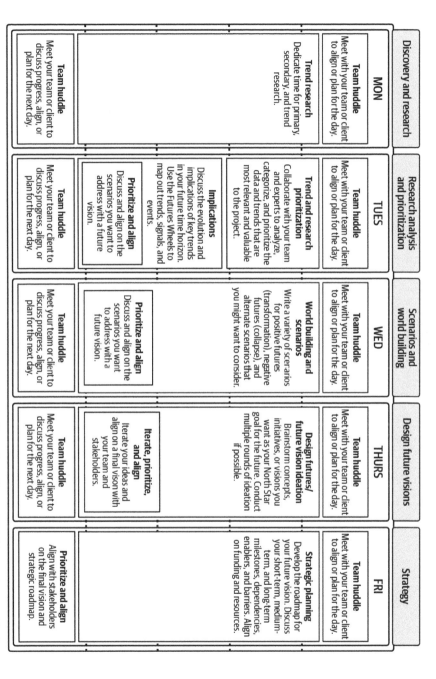

Discovery and research	Research analysis and prioritization	Scenarios and world building	Design future visions	Strategy
MON	**TUES**	**WED**	**THURS**	**FRI**
Team huddle Meet with your team or client to align or plan for the day.	**Team huddle** Meet with your team or client to align or plan for the day.	**Team huddle** Meet with your team or client to align or plan for the day.	**Team huddle** Meet with your team or client to align or plan for the day.	**Team huddle** Meet with your team or client to align or plan for the day.
Trend research Dedicate time for primary, secondary, and trend research.	**Trend and research prioritization** Collaborate with your team and experts to analyze, categorize, and prioritize the data and trends that are most relevant and valuable to the project.	**World building and scenarios** Write a variety of scenarios for positive futures (transformation), negative futures (collapse), and alternate scenarios that you might want to consider.	**Design futures/ future vision ideation** Brainstorm concepts, initiatives, or visions you want as your North Star goal for the future. Conduct multiple rounds of ideation if possible.	**Strategic planning** Develop the roadmap for your future vision. Discuss your short-term, medium-term, and long-term milestones, dependencies, enablers, and barriers. Align on funding and resources.
	Implications Discuss the evolution and implications of key trends in your future time horizon. Use the Futures Wheels to map out trends, signals, and events.			
	Prioritize and align Discuss and align on the scenarios you want to address with a future vision.	**Prioritize and align** Discuss and align on the scenarios you want to address with a future vision.	**Iterate, prioritize, and align** Iterate your ideas and align on a final vision with your team and stakeholders.	**Prioritize and align** Align with stakeholders on the final vision and strategic roadmap.
Team huddle Meet your team or client to discuss progress, align, or plan for the next day.	**Team huddle** Meet your team or client to discuss progress, align, or plan for the next day.	**Team huddle** Meet your team or client to discuss progress, align, or plan for the next day.	**Team huddle** Meet your team or client to discuss progress, align, or plan for the next day.	

Figure 6-1. A five-day futures sprint timeline. This timeline is very compressed; typically a robust Futures Thinking project would take more than a week, and you would have additional time to digest and iterate, but the scenario shown here is quite plausible.

SECURE YOUR ALLIES AND CHAMPIONS

This seems like a no-brainer, but making sure you have a good support system throughout the project is extremely valuable. It allows you to spread out the responsibility and accountability and can protect your sanity along the way. Refer to your Power-Interest Matrix (discussed in Chapter 3) if you have to try to team up with leaders who support you, whether or not they fully understand the process. This partnership can be critical when you are trying to get funding or are managing conflicts. The more believers you have around you, the less stress and weight you'll have to bear if the boat starts to rock. They can also be integral in communicating what you are doing and celebrating your wins during and after the project.

Rick Holman, an affiliate of Institute for the Future and formerly senior manager of global foresight for General Motors, discussed with me the power of seeking allies:

> I would identify your allies in the organization; there are a lot of them out there, especially in the larger organization, and they are not always in the places you might think. You want to know how many people are doing innovation as part of their jobs. They are the part of the company that prepares for the future. There are certain people for whom this work makes sense—they want to do it, they love it, they want to do it all day long. And there are other people for whom this work makes no sense at all. Don't waste your time trying to convince those people. I would identify the people in the organization, who might not have the title of foresight, but they are doing it anyway. They are your support group; they're your organization, your virtual organization, either because it's their assignment or because it's in their nature. And I would continue to build rapport with the senior leadership. One of the most successful tools is the interview. I often used interviews as a way to get them into the conversation. It is a powerful tool to create advocates.

ALIGN ON THE DELIVERABLES

Hopefully, during your diagnostic you've discovered a little more detail about what your client is expecting at the other end of the process. Maybe they want a 10-year roadmap or a speculative design of a future product. Or maybe they just want an analysis of a market in 2030 to see what competition is rising.

Try to make it clear what you are working toward so that you can make room to deliver and polish that deliverable in time. Keep in mind that the deliverable might change along the way. You may have an exciting journey of discovery and decide that a market analysis is not enough and that you want to create a "vision video" (a speculative narrative in a video format) or 3D artifact to go with it. Align with everyone on the expectations of the project. If the deliverable changes, make sure everyone knows that it has changed and that you can adapt the timeline to deliver that as well. For those instances in which your client or audience doesn't know what they want in the end, you can always show them examples of different types of outputs and have a discussion about what would best suit them for the goals they're trying to achieve. Do they need a video or website that showcases the future scenarios and vision? Or can the output just be a PDF document with pictures of the workshops, images, and text describing the work? Try to understand who the audience for the final deliverable(s) will be. Is it only internal or for public consumption? This can determine the content and type of the output that you'll need to deliver.

SET UP WORKSPACES, REPOSITORIES, AND COMMUNICATION PREFERENCES

Once the research begins, you'll be collecting various types of futures intelligence, including trends, data, articles, interviews, scenarios, and ideas. It can get overwhelming pretty quickly. Decide on how you'll manage the timeline, tasks, documents, and research and how and where you will facilitate workshops and capture outputs. Create a system that will allow you to organize, categorize, and tag your data. Once you start to collect trends, your repository can become a living document that you and your client or team can continue to use even after the project. As trends and data change over time, you can use the repository to update information as it evolves.

A few critical platforms, materials, or processes you might want to begin with are:

Administration workspaces
> Where you manage the timeline, roles, resources, activities, deliverables, progress, and status.

Research repository
> Where you collect, tag, and store data, trends, links, articles, images, videos, and so on.

Workshop workspaces

Where you facilitate workshops and where you collect the outputs. Today there are many digital whiteboards, such as Miro and Mural, that can act as a virtual workshop space and a place to collect the output.

Communication platform

The platform through which you will communicate with your team during the process. This can be a text chat group or a platform like Slack or Discord.

Templates

Documents for reporting out to your team, client, or organization. Having a set of these to deliver synthesized outputs or documentation of workshops and other activities is useful if you have to periodically report on your progress.

DETERMINE PROCESS SUCCESS METRICS

Determine how you will measure the success of the project and the output of each exercise and stage. Completing the project on time and handing over the deliverables is one metric, but you may want to have additional qualitative measurements:

- Have you uncovered new and exciting insights (threats and opportunities)?
- Is your client understanding the process?
- Does your client continue to be excited about the process and outcomes?
- Have you transformed skeptics into advocates?
- Is there excitement for implementation of the strategy?
- Are other departments or business units hearing about this project and showing interest in the process?
- Is there potential for more work after this project?

These may be soft metrics, but they can provide insight into whether the process is working well for everyone. You can claim success across many stages, but cultural and mindset transformation are important for continued support and advocacy. In Chapter 11 we'll discuss additional quantitative and qualitative metrics of success.

Organizing Your Team

Whether you are a team of one or have multiple contributors on your side and/or on the client side, you'll want to try and configure your human resources appropriately. If you are working alone, you'll need to be very strategic about how you spend your time and efforts, and you'll want to shape the activities so that you don't overload yourself trying to run every exercise to get to the finish line. But if you do have the ability to design your team, here are a few suggestions for whom to involve:

- Product or project manager
- Designers
- Strategists and analysts
- Trend researchers/analysts
- Researchers
- Domain experts
- Marketing and communications

PRODUCT OR PROJECT MANAGER

I've already mentioned the value of having a good project manager. Whether you have a dedicated project manager will depend on how the organization operates. Sometimes the position is automatically assigned, while other times you may need to advocate for someone to take this role. Whoever fills the role should have some level of experience managing projects or teams, because you may need to rely on them heavily to keep things balanced and on track. They also must be ready to help organize everything from meetings to workshops and to be a point of contact for major stakeholders or collaborators. But if you don't have a project manager, you might be the one leading the project, and that could be a learning opportunity for you as well.

DESIGNERS

There are a range of design professionals specializing in unique functions (research, UX, UI, visual, strategy, content) who are diversely equipped with skills and tools that can be helpful at every step of the way. They are trained behavioral and social scientists, anthropologists, researchers, synthesizers, and artists and can visualize patterns and ideas quickly (not that we are the only ones with these skills, but it is usually our core job). Having a trained designer on your team, no matter their role, is always useful. If you can include a visual designer or prototyper, this can also be useful for documentation or for quickly illustrating ideas.

STRATEGISTS AND ANALYSTS

If there are people with the title of *strategist* or *analyst* in the organization, try to pull them into the project. Their involvement is useful for a few reasons:

- If they are already doing this kind of work (but not utilizing a futures toolkit), it will be better to collaborate with them rather than potentially threatening their work.

- They might have powerful tools at their disposal that you can take advantage of. The magic of Excel pivot tables can come in handy for crunching numbers.

- You might learn something new about how they approach strategy.

At some consulting firms, the role of business analyst (BA) is a traditional business strategist role. Such experts are trained to gather, process, and synthesize data and deliver various types of reports and summaries to inform the greater strategy of an engagement. They are an essential part of any client service team. Analysts can do every type of research you need (quantitative and qualitative), organize and prioritize information, and help manage and direct activities if needed. They can also be powerful allies at every step of the process. However, tread lightly when working with other strategists (especially if this is a new process you are introducing into an organization), as you might threaten them or unintentionally give them the impression that you are competing with their approach. You may want to think about how you approach them. Start by recognizing their expertise, contributions, and knowledge about the organization or subject matter. Then discuss how important their position could be on the team due to the value they already bring. This approach can help reduce any perceived threat. In the best-case scenarios, strategists and analysts will want to collaborate and learn new tools and contribute toward the greater strategic outcome.

TREND RESEARCHERS/ANALYSTS

There are professionals out there whose job it is to find and analyze trends. Find them. Work with them. Get to know their craft, and collaborate with them to

gather your futures intelligence. They will have access to resources and tools that can be valuable, and they can observe and synthesize patterns quickly.

Livia Fioretti, senior trend researcher at TrendWatching, explained to me the value of having a trained trend researcher on your team:

> A good trend researcher has curiosity and a variety of interests and a willingness to leave their comfort zone. They also have a good memory and an ability to retain information so that you can connect the dots. I can curate information quickly and use systems thinking to analyze because it's a lot of information to process sometimes. We also need to be able to collaborate with others and know how to create a community to share ideas. It's important to know how to talk to people about the trends, to abstract it, simplify it, and communicate effectively.

RESEARCHERS

While trend analysts specialize in finding trends, they aren't the only ones who can do that work. A good researcher (a design researcher, an ethnographer, an anthropologist) can also find trends and other types of quantitative and qualitative data that can be useful. Trends are not the only kind of futures intelligence that is necessary for understanding the future. Thus, anyone who is trained in research can certainly contribute to the analysis of historical and present-day data and can also be trained to analyze trends. Of course, the preference would be to have a professional trend researcher and analyst.

DOMAIN EXPERTS

If you have access to domain or subject matter experts (SMEs) in the domain you're exploring, pull them in whenever possible or make them part of the team. Domain experts can provide guidance on trends, connect you with other experts, help make decisions and prioritize ideas, and give credibility to any data that you gather. Domain experts can be pulled in during critical moments in the journey. For example, you might work with an SME to gather trends and help you prioritize them. You might then pull them in toward the end of the process to help you ideate on future visions and scenarios. They may not have to be part of the entire process but instead might participate only in moments at which their expertise could be critical to the outcome and to provide validation that an expert did help shape some of the thinking and ideas. This credibility check can be useful when having to report out to stakeholders or other key leaders who might scrutinize the work.

MARKETING AND COMMUNICATIONS

I like to make sure the marketing and communications department (aka Mar-Comm) is aware of the project at some point. I discovered this while working on futures projects and planning large events. MarComm departments can have connections, influence, and impact across an entire organization. They can connect you with people and evangelize your work, and you can provide them with inspirational content to communicate internally and externally. In 1987, Apple Computer created a vision video about a speculative product called the *Knowledge Navigator* (Figure 6-2), a foldable, portable computing device with features like an AI assistant, a touchscreen, and wireless communication—all of which did not exist yet. The video became a huge success during its time as both a marketing tool and a vision statement about where Apple wanted to go. Because MarComm is usually responsible for creating imagery for announcements or organizing events, they may have access to design and production resources that you might need for some of your outputs. They can also help with distribution and exposure to the public or to other networks you might want to tap into.

Figure 6-2. Apple Knowledge Navigator 1987 (https://oreil.ly/OQ3iI). Conceived by then–Apple CEO John Scully; written and developed by Hugh Dubberly, Doris Mitsch, and Mike Liebhold. The Navigator was initially created as a vision for future Apple products but included such modern tech as a touchscreen foldable tablet, voice assistant, and video conferencing, and alluded to communication infrastructures such as WiFi and the internet (neither of which were commercially available at the time).

MEMBERS VERSUS COLLABORATORS VERSUS ALLIES

Across any of the roles previously mentioned, it will also be useful to distinguish between team members, contributors/collaborators, and allies, as how they participate could tell you how much and to what degree you'll be needing them throughout. This list of suggested team members and collaborators is not exhaustive; your team could include other roles such as developers, engineers, legal teams, and finance. There are essentially three different types of people you can include:

Members

These are the core team members. In a consulting practice, you may have members on your consulting team and members of the client team. In some configurations you and the client team work side by side on the project, whereas in other configurations you engage with the client team only during certain ceremonies or meetings for input, collaboration, or report-outs. For internal work, you may have a core team, but you may also have people who are provisionally part of the team but are not full-time on the project. These are the futures lead, the product/project manager, designers, researchers, and strategists.

Contributors/collaborators

The people in these roles may or may not be part of the official team but are contributing or collaborating throughout the process nonetheless. They can include domain or subject matter experts, engineers, analysts, or employees in other business units who are not on the team but are submitting trends or inputs.

Allies/champions

These are people who are not officially part of the team but who can be essential in supporting you financially or operationally. They can include executive leaders and members of the marketing and communications, finance, and other departments who may support or have interest in the work but are not directly connected to the team.

Depending on the context of the project and when you need to involve them (for instance, finance may need to be involved only at the beginning and at the end), you may have to decide with your team who is critical to have as part of the core team and who is ancillary or can be on call as needed throughout the project. While it's important to be inclusive and have many voices contributing throughout the process, too many team members can also create challenges in managing schedules and commitments. But do consider having a strong and diverse core team (it's also important to make sure there is ample representation from different genders and cultural backgrounds, if possible) that will be dedicated to the process from the beginning to the end so that there is accountability throughout.

In Closing

While *problem* framing is like drawing a map and planning the route to your uncharted island, *project* framing is like making sure your boat is equipped and ready to sail—it's a necessity if you don't want to sink or die in open water. Attending to these details will keep you from boiling the ocean throughout the project. Project framing also keeps everything within a defined scope so that you can accurately balance budget, time, and resources. By now, you might have a full agenda and timeline planned. All the work from Stage o should be cohering into a solid plan for beginning the futures work (see the following sidebar). Funding, time, and allies are secured. You know what your client or team wants. They are excited and have adopted a futures mindset. Now we're ready to start sailing into the future. In the next chapter, I'll discuss Strategic Foresight and how to use it to start gathering intelligence and data about the past and present so you can start imagining future worlds.

Checklist for Stage 0

1. Diagnostic: Understand your audience

   — You've used various situational awareness methods or frameworks to understand your audience, team, or organization: the projects, people, and environment you'll be futuring with.

   — You understand now how much you have to prepare them for the process and journey.

   — You have a sense of where they are today (current state), where they want to go (future), and how you might want to take them there (the stages or methods you want to use).

2. The Futures Mindset

   — You have prepared them for futures work by showing them examples and explained the principles and process.

   — Everyone has a clear idea of the approach. They're excited and ready to be led into the future.

3. Problem Framing

   — From your diagnostic and inquiries, you have determined and aligned on the detailed parameters of the project: the focal issue, time horizon, geographic boundaries, demographics, complexities, and enablers.

   — You have aligned on the outputs you need for the project.

4. Project Framing

   — You've set up the team, timeline, funding, and logistics.

   — You are all aligned on the deliverables and stages of the process.

   — You have organized repositories, project management tools, and documents.

Checklist for Stage 0

1. Diagnostic: Understand your audience
   - You've used various situational awareness methods or frameworks to understand your audience, team, or organization: the projects, people, and environment you'll be futuring with.
   - You understand now how much you have to prepare them for the process and journey.
   - You have a sense of where they are today (current state), where they want to go (future), and how you might want to take them there (the stages or methods you want to use).

2. The Futures Mindset
   - You have prepared them for futures work by showing them examples and explained the principles and process.
   - Everyone has a clear idea of the approach. They're excited and ready to be led into the future.

3. Problem Framing
   - From your diagnostic and inquiries, you have determined and aligned on the detailed parameters of the project: the focal issue, time horizon, geographic boundaries, demographics, complexities, and enablers.
   - You have aligned on the outputs you need for the project.

4. Project Framing
   - You've set up the team, timeline, funding, and logistics.
   - You are all aligned on the deliverables and stages of the process.
   - You have organized repositories, project management tools, and documents.

Strategic Foresight

This stage (the first stage of actual futures work) introduces Strategic Foresight, one of the cornerstones of traditional futures work. As noted in the introduction to this book, foresight tools are powerful in the research stage of futures work. Strategic Foresight allows you to understand what you want to investigate that eventually influences future environments and to develop a perspective on multiple scenarios so that you can prepare for future threats or opportunities.

In this stage, we'll explore the following:

- Chapter 7, "Fundamentals of Foresight"
  - Working with trends
    - What are trends and the drivers that create them?
    - How to identify trends and categorize them based on size and impact
  - Analyzing and prioritizing trends
    - How to determine which trends to focus on
  - Implications
    - How to look at the implications of trends, events, or signals and how they might evolve, change, or impact other trends, sectors, or areas of interest
- Chapter 8, "World Building with Scenarios"
  - The history and purpose of scenarios in foresight
  - How to use research and trends to formulate scenarios
  - The different types and formats you can create

Fundamentals of Foresight

In this chapter I'll discuss some of the fundamental theories, principles, and methods used in Strategic Foresight, a discipline that has been around since the 1950s and that has been used primarily for scenario planning and long-term strategic initiatives. Foresight is a form of futures work and has its own collection of tools and frameworks for analyzing patterns around us so that we can project a perspective of possible future scenarios.

In his 1999 book, *Futures for the Third Millennium: Enabling the Forward View* (Richmond), Richard Slaughter describes Strategic Foresight as

> *the ability to create and maintain a high-quality, coherent and functional forward view, and to use the insights arising in useful organizational ways. For example to detect adverse conditions, guide policy, shape strategy, and to explore new markets, products and services. It represents a fusion of futures methods with those of strategic management.*

Strategic Foresight

In a traditional design-driven approach (such as Design Thinking or Service Design), you might consider as this similar to the first half of the first diamond in the Design Council's Double Diamond diagram (*https://oreil.ly/3faby*) (Figure 7-1), the *Discover* phase, which illustrates the divergent research phase. In foresight, we use research tactics to understand the trends, patterns, history, and current perspectives of today. We gather as much information as possible to fully understand the scope of the problem space. Then we synthesize, prioritize, and converge, just as we do in the convergent *Define* phase of the Double Diamond, except we are converging on scenarios and implications of the future, projecting and building future worlds so that we can understand them better, with more focus on specific areas of exploration.

If we want to utilize Futures Thinking to design future products, services, or initiatives, we can visualize the process as three diamonds in which each phase is diverging and converging (Figure 7-2). In the first diamond, we are doing traditional research, using a variety of tools and looking for trends and patterns. That exercise can quickly become overwhelming, so we must converge and prioritize the trends and research that we want to focus on. In the next diamond, we diverge by creating multiple scenarios and discussing the implications of the trends and research we've synthesized. This allows us to build alternative possible futures. But again, there could be myriad possibilities, so we would have to prioritize and select which worlds we want to address or design for. Which are the most imperative scenarios and implications, the ones that matter most to our business, our strategy, our bottom line, or the health of our society? There may be one scenario or multiple scenarios, but we can't solve for every one, so we would prioritize. From there, the last diamond is similar to the second diamond of the original Double Diamond, where we can begin to employ traditional design ideation methods to generate multiple ideas—and we would again have to select and decide which ideas are the most important or most valuable to designate as our North Star initiative to strategize around.

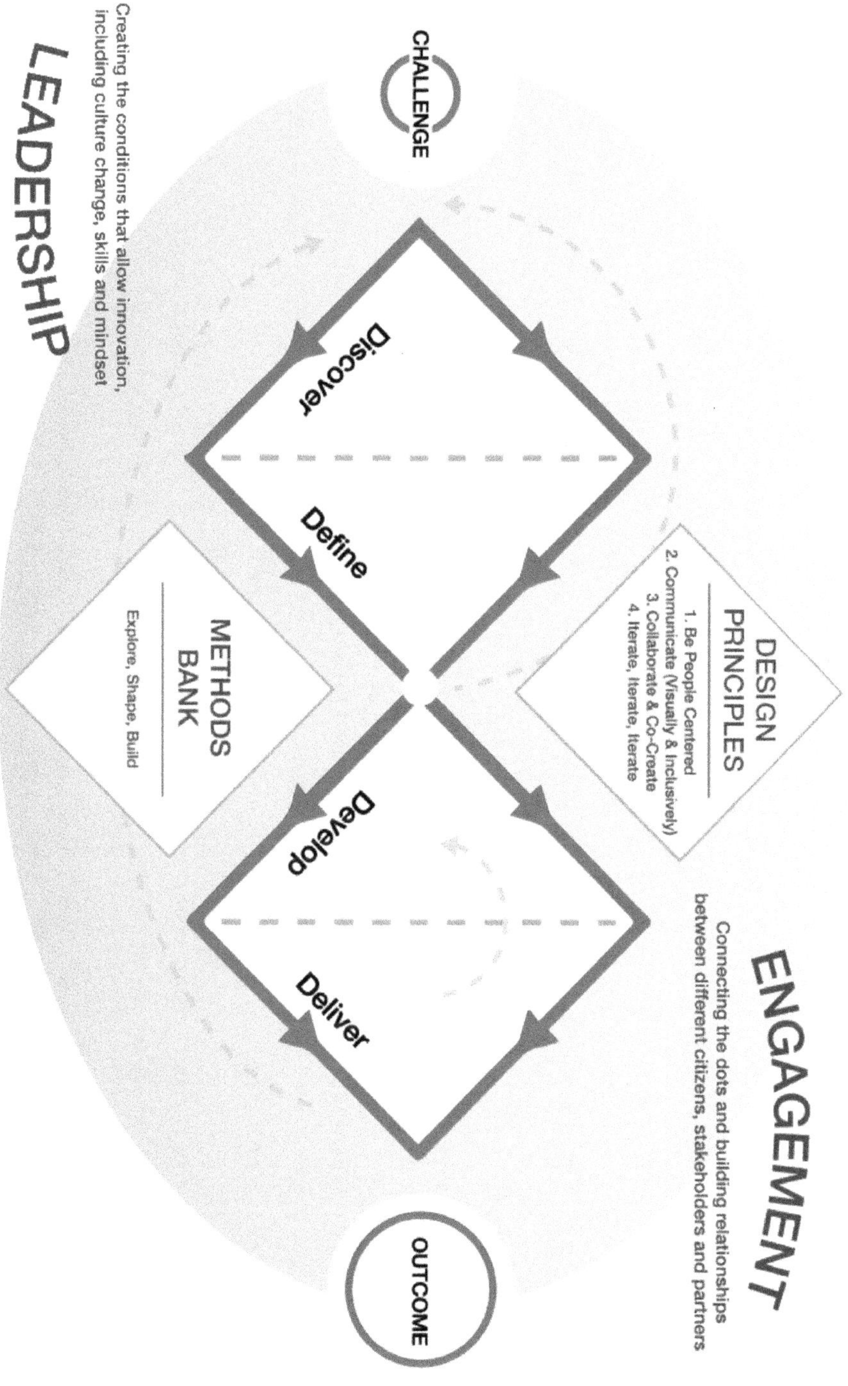

CHALLENGE

Discover

Define

Develop

Deliver

OUTCOME

LEADERSHIP

Creating the conditions that allow innovation, including culture change, skills and mindset

METHODS BANK

Explore, Shape, Build

DESIGN PRINCIPLES

1. Be People Centered
2. Communicate (Visually & Inclusively)
3. Collaborate & Co-Create
4. Iterate, Iterate, Iterate

ENGAGEMENT

Connecting the dots and building relationships between different citizens, stakeholders and partners

Figure 7-1. The Double Diamond (source: the Design Council)

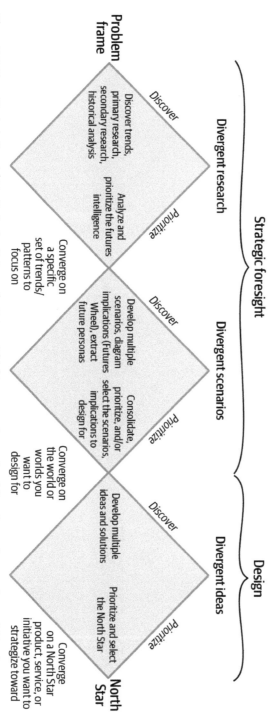

Figure 7-2. From Strategic Foresight to Designing Futures, the Futures Thinking process is a series of diverging and converging activities that may ultimately lead toward a North Star or future vision goal

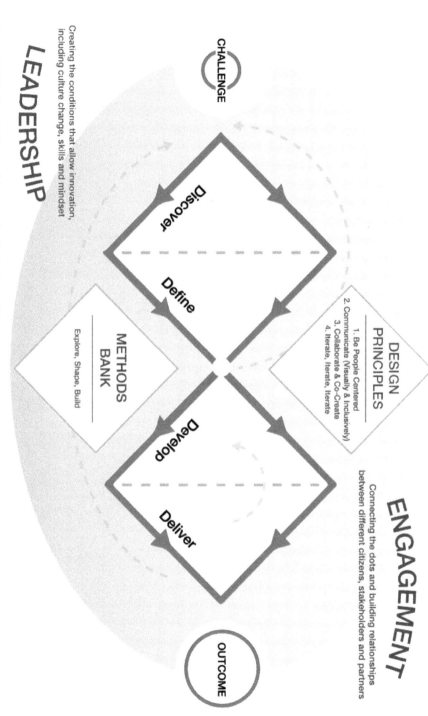

Figure 7-1. The Double Diamond (source: the Design Council)

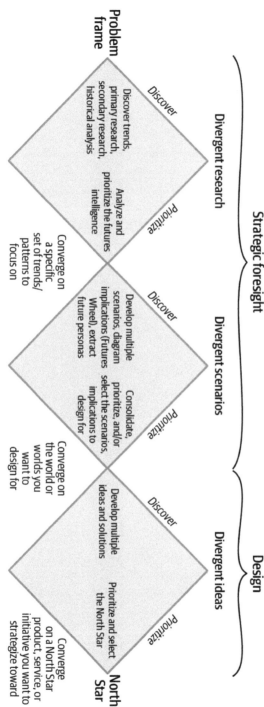

Figure 7-2. From Strategic Foresight to Designing Futures, the Futures Thinking process is a series of diverging and converging activities that may ultimately lead toward a North Star or future vision goal

In the research phase, we also conduct primary research tactics (user interviews) and secondary research tactics (desktop research) to investigate everything we want to know about our problem space (focal issue). We do this by scanning the future horizon for signals of change. These can be patterns and trends we see in the world. Trends are patterns (multiple instances of an idea or change that have commonalities) or shifts that have scaled in popularity and exist in many places. They can be large patterns (like everyone moving out of cities due to inflation) or small trends (like a group of friends who pay only in Bitcoin). I'll discuss this in greater detail later, but trends are one of the core categories of research we investigate to understand what the future could be like.

Some of these steps may seem familiar; that's because they are very similar to traditional design research. We talk to people, we gather insights, we cite papers, books, podcasts, and thought leaders. The only difference is in the way we process that information and how we use it. The mental models are the same in that we need to collect, aggregate, and make sense of everything we find, but the tools and analyses are a little different in that we project that information into futures so we can start to understand how those patterns may play out over time.

One difference you may find between Strategic Foresight and traditional design research is that you are not necessarily trying to solve an immediate problem or solve for one persona but instead are trying to gather information about a future point in time so that you can assess whether or how you might want to solve for something that may not exist yet. Some might ask, "Why would I want to spend time thinking about problems that don't exist yet? Shouldn't we be solving the problems of today so that we can survive?" These are valid questions, but they are not necessarily the best questions to ask if you want to understand the value of futures work. I will make arguments about this throughout the book. We'll begin to see that we aren't necessarily ignoring today's problems; we're just trying to stay ahead of them, be proactive, and set long-term goals that also inform how we can solve today's issues by setting up more informed and responsible perspectives and strategies toward the future. Return on investment (ROI) is also something I'm constantly asked about with regard to futures work. There definitely are ways to show how futures work produces tangible value, though they might not look like our traditional near-term metrics, because they also incorporate "soft" aspects such as changes in attitudes, beliefs, and culture mindsets. We'll delve further into how to measure ROI in Chapter 11, which discusses designing a futures strategy and closing the loop on futures work.

As I discuss some of the core principles and methods of Strategic Foresight, you might wonder whether it's necessary to use all these methods to properly do futures work. The short answer is no. But it will serve you well to try out what methods work for you and your projects, as they offer a variety of ways to analyze and synthesize information and can generate a multiplicity of details for how you might want to think about future worlds. If you worked on a Design Thinking project and decided to omit an empathy map or use a service blueprint as opposed to a standard customer journey map, no one would shame you for it, as long as your approach and solution met the requirements and your users' needs, right? The same mentality can be used when working with a futures toolkit. Use what you need within the time you're given to provide the outcome you are looking for.

World Building

One of the intentions of Strategic Foresight is to illustrate potential futures or future worlds. We do this through a series of exercises that allow us to slowly construct the characteristics of what the future might look like. Sometimes we refer to this as *world building*—a term typically used in literature, where you are literally *building* a world for the reader to journey through. In literature, readers are slowly introduced to the actors in the narrative, the rules and laws of the world they live in, and the context of the story. In futures, each world we build also has its actors, rules, and context. We can use the information we gather from foresight activities to build that world for our client or audience. Each world can have different technologies, conditions, dependencies, and temperaments. We call these *scenarios* (which I'll cover in more detail in the next chapter). One world may be dystopian and have a collapsing economy filled with turmoil, while another might be transformational and rich with innovation, growth, and progress. We want to stop and look around inside these worlds so we can determine how we might want to operate in them, what products and services we may want to build, and/or how these environments could affect our lives, work, and strategies today.

World building is useful in establishing the belief system necessary to investigate future scenarios. If we said "now you're in the future" and didn't tell you much about it, you might think the future looks similar to today (because that's all you know) and design something that meets only today's needs, ignoring all the conditions in the world that could change over time. But if we were to give you more detail about culture, technology, politics, and other trends that could reshape civilization, we would be having a totally different conversation. The information that shapes your belief system allows you to navigate any world with more knowledge, confidence, and curiosity, allowing for much deeper conversations about the topic you're investigating. People who have strong religious beliefs (and I'm not saying that futures is a religion per se) rely on their deities and the principles of their religious texts to guide the way they live their lives. Their belief is so strong that it can influence their values, which in turn drives the decisions and pathways in their personal lives and careers. Thus, to think about the future, we have to believe the world *could* exist whether or not we have substantial evidence. We rely on information we have and even on trust and faith (even if it's temporary) so that it can drive our decisions today. Without a belief system, you nurture a skepticism that can make it difficult to determine what might impact you or your design strategies. Without a belief system in place, your work and aspirations will fall on deaf ears.

In the early 1960s, the US was perceived as losing the space race to the Soviet Union, which had launched the first artificial satellite in 1957. On May 25, 1961, President John F. Kennedy proposed that the US "should commit itself to achieving the goal, before this decade is out, of landing a man on the moon and returning him safely to the earth." The following year, on September 12, 1962, President Kennedy delivered a speech to a crowd of about 40,000 people at Rice University's Rice Stadium that became famous as a rallying cry to place a man on the moon:

> We set sail on this new sea because there is new knowledge to be gained, and new rights to be won, and they must be won and used for the progress of all people. For space science, like nuclear science and all technology, has no conscience of its own. Whether it will become a force for good or ill depends on man, and only if the United States occupies a position of pre-eminence can we help decide whether this new ocean will be a sea of peace or a new terrifying theater of war. I do not say that we should or will go unprotected against the hostile misuse of space any more than we go unprotected against the hostile use of land or sea, but I do say that space can be explored and mastered without feeding the fires of war, without repeating the mistakes that man has made in extending his writ around this globe of ours. There is no strife, no prejudice, no national conflict in outer space as yet. Its hazards are hostile to us all. Its conquest deserves the best of all mankind, and its opportunity for peaceful cooperation may never come again. But why, some say, the moon? Why choose this as

our goal? And they may well ask, why climb the highest mountain? Why, 35 years ago, fly the Atlantic?...We choose to go to the moon. We choose to go to the moon in this decade and do the other things, not because they are easy, but because they are hard; because that goal will serve to organize and measure the best of our energies and skills, because that challenge is one that we are willing to accept, one we are unwilling to postpone, and one we intend to win, and the others, too.

Kennedy's speech was not only a declaration and a challenge but also a motivation for the US and NASA to accomplish what up to that point had been unachievable. It was also a political effort to surpass the USSR during the Cold War and instantiate the US as a stronger world power with a mastery over space sciences. It was a call to action that resonated with many, setting the foundation for the Mercury program (to determine whether humans could survive in space), the Gemini program (to test maneuvering in space), and the Apollo program, which culminated in landing a manned spaceship on the moon on July 20, 1969, at 20:17 UTC (Figure 7-3). Kennedy's speech was a legendary form of rhetoric that established a belief system. Within it, Kennedy used three strategies: "a characterization of space as a beckoning frontier; an articulation of time that locates the endeavor within a historical moment of urgency and plausibility; and a final, cumulative strategy that invites audience members to live up to their pioneering heritage by going to the moon."[1] Embedded in the speech was a narrative of potential threat illustrating details and dramatizing the plausible future that could arrive, but there also was a message of hope and prosperity for Americans and our species. These are merely scenarios (which we'll dive into more in the next chapter), used to describe a world we would want to avoid and one we need to create to prosper. Used properly, rhetoric, storytelling, and imagery can be powerful tools to activate the hearts and minds of people, and this is precisely how futures and foresight can be used to incite action today. Today, Kennedy's Rice speech is still evoked when people talk about a *moonshot* project—a monumentally challenging and innovative project or undertaking. Moonshots are lofty goals that seem almost impossibly out of reach but may be attainable through rigorous effort and problem solving. Google X (now just

---

1 John W. Jordan, "Kennedy's Romantic Moon and Its Rhetorical Legacy for Space Exploration" (*https://oreil.ly/iRc8W*), *Rhetoric and Public Affairs* 6, no. 2 (Summer 2003): 209–231.

called X) was a moonshot factory and meant to be a test bed and incubator for experimental ideas that were extremely ambitious, such as delivering internet access to remote areas via high-altitude balloons.

Figure 7-3. Moon landing, July 20, 1969: Buzz Aldrin stands next to the Passive Seismic Experiment Package, with the Lunar Module, "Eagle," in the background (source: NASA (https://oreil.ly/-Ctbu))

HORIZON SCANNING

One of the first activities in gathering intelligence about the future is *horizon scanning*. As the name indicates, this is the act of scanning the horizon for clues about what's to come. We do this by seeking various sources around us to identify patterns and trends. Some believe that trends are the lifeblood of foresight work, and while the research phase does culminate in and utilize trends, we can also tap into many other sources of intelligence (which I'll explain in the following section about trends and gathering futures intelligence). But the future isn't just about what we see *today* projected forward in time—we also have a wealth of data in our past. Scanning backward into our history can also be very useful in understanding how patterns emerge, evolve, transform, or merge into other patterns.

Shaping Tomorrow, a foresight platform for education and trend scanning, defines horizon scanning (*https://oreil.ly/A6tXZ*) as

> *a technique to detect and identify emerging developments, trends and other signals of change through the systematic analysis of trusted source materials. It is used by businesses, organizations and governments to identify risks and opportunities in their emerging environments. A horizon scan provides space within which appropriate action can be taken before the full impacts of changes are felt. The information gathered during horizon scanning can be essential when creating strategy or responding to these changes, such as natural disasters, political instability, unstable market forces, etc.*

Looking for patterns and trends in the world is much like playing a sport: mastering it takes a lot of practice. But once you've done so, it will allow you to read between the lines of an article or a report and translate the information into identifiable patterns. This not only contributes to the wider act of knowledge gathering but also builds futures literacy, new perspectives, and other capabilities that fuel the futures process and cultural mindset shifts necessary for looking into the future. However, scanning doesn't just involve looking for patterns in the world today; it also requires an ability to look laterally across sectors (not just your own) so that you can see oncoming threats or influencing forces.

Livia Fioretti, Senior Trend Researcher at TrendWatching, says this about building a constant habit of trend scanning:

> *If you really want to do trend research, I think it's important to change the mindset that this is not something you just do for a specific project that you need to deliver for a report or to present something in class. It's really important to think of it as a training within your projects. There really isn't a beginning and an end. It should be something that's always evolving, because trends can go in completely different directions out of the blue. So my approach is for everyone to be doing it, even if it's from a very amateur perspective. Have a Miro board or a notebook or Notion page where you can just add your notes and keep adding to it.[2] Because that's the only way you can see the evolution. And it doesn't need to be anything complex. It can just be a series of notes, but it's important to keep doing it because the problem we find when we search for trends is it is so time-dependent, and if you don't keep on zooming in and out, it will grow old and all the research will be lost. So I feel like the value of it is the consistency.*

HISTORICAL ANALYSIS

The word *futures* may inherently suggest looking forward, but horizon scanning in futures work is really built on what we know about the present and the past. Tapping into the past is a useful way of discussing drivers of change that have perpetuated into patterns we see now. We can call on the past to determine how humans behaved or responded when faced with certain events, or how larger or other systems responded. Analyzing our success and failures should be innate, but you'd be surprised how many people forget that we may have been here before and that we shouldn't be making the same mistakes.

If we're interested in how a particular trend, event, or emerging technology could impact our future, we may want to see if there is anything in our history that is similar that we can learn from. When was the last time a new technology was introduced? How did customers respond? What were their concerns, and why did the technology succeed or fail? You could look at how specific events, generational events, or past global trends reshaped economies. Larger movements such as the Fourth Industrial Revolution (*https://oreil.ly/udfmF*) can

---

2 Miro is a digital whiteboard; Notion is a collaborative productivity tool.

tell us a lot about how companies were formed or collapsed, about policy issues or ethics that needed to be considered, or about the greater impact on human behaviors and society. Every year, as technology evolves and integrates faster into society, we learn more about what we believe is good or bad and are in a constant race to try and control what we are doing to avoid doing more harm to the world or ourselves. So why aren't we referencing historical events more often as examples of what not to do? For instance, during the recent COVID-19 global pandemic, many historians and economists looked back to the Spanish flu pandemic that occurred in the early 1900s (Figure 7-4). Education was severely impacted; children who were born during the pandemic or suffered through lockdowns were deprived of traditional education. Students were forced to find alternative methods and places for education, as occurred during the COVID-19 pandemic as well (Figure 7-5). The number of graduating students dropped.

Figure 7-4. May 1919: Students in their "backyard workshop" in Denver, Colorado, while school was closed during the Spanish flu pandemic (source: Universal History Archive/Universal Images Group/Getty Images)

Figure 7-5. April 2020: Homeschooling during the coronavirus pandemic (source: weforum.org (https://oreil.ly/ktN1g))

The ability to gather accurate data was not as advanced in the early 1900s as it is today, but we were able to record the effects from various records, including economic impacts globally. While the world was much different a hundred years ago, there were patterns in how the government, society, and global economies reacted to the flu pandemic's effects. We saw a downturn in spending, and supply chains and government resources were challenged and experienced failures. Finances were bottled up for years as economies ground to a halt. These same events occurred in 2020–2021. And as society regained control of the Spanish flu and began to recover, a surge in spending bloated the US and global economies, and a recession followed. We learned a lot from those events, and we saw how history repeated itself this century, as a recession proliferated into 2022 and 2023. There were massive tech layoffs, a war in the Ukraine challenged energy security, and inflation continued to surge in many countries. While these events may seem new and unexpected to some in this current era, the patterns are not and were anticipated by many analysts, futurists, and economists. Yet they still seemed to surprise most people.

When we scan horizons from the past and present and into the future, we are also looking for shifts in culture, innovation, politics, and economies to try to understand how they will shape the future. While some patterns are new and emerging and can still be affected by anomalies or hidden forces that we might miss or disregard, we can marry what we know about the past to what is happening today and then try to estimate how some patterns will play out. Again, this is a game of possibilities, with an amorphous and dynamically changing target. Things can always play out differently, but by constantly scanning around us, we can accurately anticipate some events.

BILL SHARPE'S THREE HORIZONS

Much like we use the Futures Cone to explain alternate futures and the Futures Triangle to explain influencing forces in the world, there are mindsets and frameworks in foresight that can help us understand change over time from different perspectives. With these frameworks, we can contextualize certain trends or patterns, identify how they may disappear or scale, and use them to find areas to innovate, disrupt, or intervene in with strategies. Bill Sharpe's Three Horizons (Figure 7-6) is a framework that looks at the future in terms of three different perspectives. Sharpe has described Three Horizons (*https://oreil.ly/tE2JW*) as "a way to think about the future that recognises deep uncertainty but responds with an active orientation; that allows us to understand more clearly how our actions and those of others we engage with might shape the future we are trying to explore." He portrays it as a passionate conversation in which there is a voice of concern (I can't really see things improving), a voice of enterprise or entrepreneurship (there must be something we can try), and a visionary voice of aspiration (if we can imagine this world, then surely we can build it); these voices correspond with each horizon:

Horizon 1 (H1)

This is the voice of concern, representing how we do things today. We are all part of a system, and by acting within that system we perpetuate how the system works and feed it what it needs. But we all know that things change and that the system will eventually not be enough. As needs and opportunities emerge or some new technology or process arises, the system or some facets of it will become obsolete.

Horizon 2 (H2)

This is the voice of enterprise; it represents the innovation that rises to meet H1 and H3, incorporating new needs and changes in society and the world by bringing new technology and processes to reinvent or replace the H1 ways of doing things.

Horizon 3 (H3)

This is the voice of aspiration—a vision of aspiration, transformation, and what the future could bring. It is composed of changing societal values, opportunities, and needs. It represents an evolution of thought and a rediscovery of the world and how we want it to be. It emerges and transcends our current beliefs about how systems of today work, and challenges the status quo.

According to Sharpe, all horizons are conversations we are constantly having with ourselves and others every day. There are those who adhere only to H1 thinking and don't think change is necessary (business as usual), and those who are extremely visionary and are reading the patterns in the world and aspiring to new H3 futures. H2 voices may be able to understand both perspectives and seek to solve H1 inefficiencies with H3 visions.

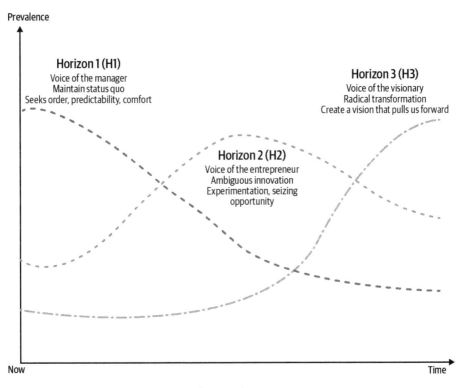

Figure 7-6. Bill Sharpe's Three Horizons framework

The Three Horizons framework is a way of understanding different perspectives and trends over time and how you might manage these "conversations" or view patterns as they develop. There is no right or wrong horizon to align with, but by identifying how people or businesses think about the future, this framework can help you become a voice that can influence conversations and, ultimately, organizational change.

For example, you might have parents who have always owned American-made gas-powered cars (they may even prefer a certain brand). They might not want to try hybrid or electric cars at all. Their preference could be influenced by their experience of the reliability of the car they've used for years, politics (they might lean conservative and associate electric cars with liberal thinking), or even friends' opinions. The parents could be considered H1 voices. You, on the other hand, might be more eco-conscious and may truly believe in the sustainability of a fully electric vehicle. You might be the H3 visionary who sees the future and the positive potential of moving society away from fossil-fueled vehicles and into other renewable sources. Your entire life and ecosystem of products could be driven by your views on environmental concerns. Your older sibling, however, might be somewhere in the middle. They own a hybrid because it still has the range of a traditional gas-powered vehicle but also saves money and the environment. They may try to convince your parents to invest in hybrids or campaign for your parents to eventually detach themselves from their dependency on their older vehicle. Your sibling sees the future and is able to invest in it; they also recognize the obsolescence of current dependencies and might facilitate the conversation to change the H1 parents' minds. Maybe they won't fully transform their politics or friends' opinions, but the H2 mindset seeks to accept realities that might not have a place in the future and seeks ways to bring transformation and clarity to some conversations.

Three Horizons can also be used when explaining the evolution of technologies or processes. H1 horizons can represent fossil-fuel technology in transportation, which hopefully will degrade into obsolescence as we continue to innovate and replace it with new renewable H3 technologies like hydrogen, solar, or other alternative fuel innovations. We are in the midst of observing this horizon, but it can take many decades for the H1 to fully deprecate from the world. H2 horizons can be seen as the hybrid vehicles that still use fossil fuels but incorporate electric drive trains. As the technology and fuels continue to develop, we will see deprecation of certain hardware, manufacturing techniques, and fuel resources and the rise of new processes over time.

According to Rick Holman, an affiliate of Institute for the Future and formerly senior manager of global foresight for General Motors:

When you're looking out into the future and talking to people who are inventing and creating it, they're also creating new languages. They're creating not only the technology, and thinking about how it's going to affect us, but they're also thinking about it in ways that aren't popular or mainstream yet, so they're kind of inventing a new language. One of the big roles of a foresight person is to translate the ideas and the new language back into a language that the organization can relate to. If you didn't do it well, you're going to be looked at as weird. So a big part of the foresight role is contextualizing. I always viewed it as having a foot in two worlds; on the one hand, you are deeply part of the organization, but you don't want to be consumed or assimilated into the group think in such a way that you aren't as open to ideas. I'm sure, as you've experienced, when you get really into a culture, it's really easy to dismiss ideas that challenge the status quo or challenge your position. That's what you're always up against.

In a big company, there are certain orthodoxies, and a good futurist challenges those orthodoxies, gets excited when they see something that challenges an orthodoxy, wants to pursue it, wants to understand it, and sees it as an opening for a possible new future. But you can't come back with that excitement without doing some translating and understanding what the decision makers are thinking, what their mindset is and what they're thinking about. It has a lot to do with being excited about and being aware of the potential dangers of the future and having that energy around that, but also having a deep respect for the culture and the direction of the company are almost certain limitations—blind spots that the strong culture and especially strong cultural successes provide. Honor those blind spots without being too critical and reframe what you're learning in a way that gets people excited, rather than triggering the immune system.

FUTURES INTELLIGENCE

Research in foresight is really an act of gathering intelligence from the world around you. So before we talk about trends, which some think are the only kind of research you do in foresight or futures work, let's talk about other kinds of data you can gather besides trend reports. We generally call this *futures intelligence*, which refers to all the information you can gather to inform your future scenarios and worlds. If you have a fairly descriptive focal issue, then you already have some direction for where you want to start your research. For example, if you wanted to look at "the future of transportation in California in 2030," you already have search terms to plug into Google or ChatGPT, and you might find several articles or reports right away. Or you may cast a wider net and just search for "the future of transportation"; given that transportation is a pretty broad and popular topic, this is sure to return some articles by people who have already done this work. It's a good starting point, but you need more information. You may now want to expand your search to include the region and time horizon. While Google is a great place to start, do remember that there are many sources and repositories out there that might have information you need. You may want to initiate a few interviews with domain experts or citizens within a region to get a wider viewpoint of public opinion. Domain experts are a great asset because they can point you to more specific resources or articles. It's likely they are already futuring in their head and anticipating change within their market so that they can better consult for those who need it. Other sources of intelligence are:

- Academic whitepapers
- Books/audiobooks
- Podcasts
- Conference talks
- News articles
- Social media (X [formerly Twitter], Facebook, TikTok, Reddit)
- Trend reports

It will take some practice scanning and digesting information from these sources to understand what is a pattern or a trend, but as you immerse yourself in these resources, looking for specific clues associated with your problem frame and focal issue, you'll start to recognize the information you are looking for. Just doing searches on the topics in your social media feeds can bring up a lot of information, but learn to scan quickly and file away articles, videos, or reels as you see them so you can parse them later to determine what might be significant.

As we noted in Chapter 6, having a repository for links and documents will prove useful as your body of knowledge grows. Scan anything and everything you have the time and budget for, and start taking notes of patterns you see. If people are mentioning a particular trend, company, idea, or technology, tag it. Chances are that as you read and discuss this topic with people, you will start to become the expert yourself and will be able to replay what you learned with your client when it comes time to review your research.

TrendWatching's Trend Hunting Framework

Livia Fioretti, Director of Insights at TrendWatching, uses a framework (Figure 7-7) based on three pillars: basic needs (human or society), drivers of change (patterns or trends that are driving innovations), and innovations (advancements or evolutions in technology).

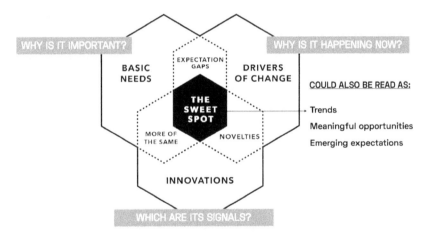

Figure 7-7. TrendWatching's trend hunting framework

Basic needs incorporate human, societal, or even species needs. You can think about this in terms of Maslow's hierarchy of needs (*https:// oreil.ly/dXA7D*) (Figure 7-8). No matter what era we are in, all humans have needs. Those needs will be the same in the future, but they will be enabled, facilitated, or met by future technologies. Drivers of change can incorporate anything from trends across social/cultural, economic, environmental, political, or ethical perspectives to other patterns that cause momentum.

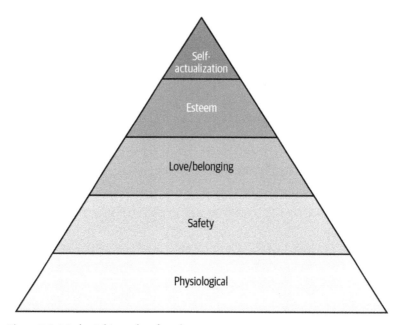

Figure 7-8. Maslow's hierarchy of needs

As Livia explained to me:

Drivers of change refer to a wide range of factors that can influence society and consumer behavior. This includes not just what people buy, but also the information they absorb and how they do so. It encompasses everything from cultural influences to social media and other forms of consumption.

Innovations are the new technologies, processes, or services that are introduced into the world. If a technology serves a human need but doesn't listen to or respond to other drivers of change, then it only serves more of the same. If a technology is responding to a driver of change but not a human need, then it disregards what we inherently need to survive and be happy. But when an innovation addresses both human needs and drivers of change, then there is a sweet spot in which we can see innovation and trends being created.

Let's use micromobility scooter rentals as an example. Electric scooters had existed for several years before they became popular. Innovations in the technology allowed the batteries to last longer and made them smaller and lighter and less expensive to manufacture at scale. Fast, accessible, fun, and affordable transportation was a *human need*, especially in cities. *Drivers of change* were the ride-sharing service trends. Sharing services by the 2020s were huge businesses, and many were already getting into the bike share business. The introduction of an even more lightweight personal vehicle that people could rent to get across town met both the human need and the driver of change (trend) that had already made businesses like Uber and Lyft popular. But convenience was also a factor—the ability for customers to just pick up a scooter and leave it wherever they wanted added an extra layer of human need (convenience) that allowed scooter-rental and bike-rental businesses to boom. The innovation of micromobility services transformed shared mobility in terms of business models, vehicles, and customer experience. If you look closely at how these three areas intersect, you might find more trends that could drive innovation across transportation, delivery services, and even community building.

Livia's framework seeks to understand not only how forces influence our needs and how technology can play a role in identifying trends and new patterns in the world, but also how we can logically understand what people want and how the forces around us can influence those innovation ideas.

USING STEEP AS A MULTIVARIATE LENS

When gathering trends and other futures intelligence, you'll naturally want to organize your research so that you can synthesize it more effectively. A useful method for scanning for trends and for organizing your research is called *STEEP*, which stands for *social, technological, environmental, economic,* and *political* factors in a given environment. STEEP is extremely useful as a research lens, a categorization tool, or a way to investigate a topic from different angles or different moments in time. We can look at the STEEP of a system *today* and, depending on what focal issue we are investigating, have a pretty good understanding of social conditions, the technology available, and popular political beliefs. If we are looking at STEEP of the *future* in the year 2030, we could extract and analyze the trends that are driving each category today and push them forward into 2030 to determine how those conditions might change or transform over time. STEEP essentially allows us to take a slice of different dimensions of a world so that we can organize our research more holistically. It can serve as a good starting point and a way to make sure you are considering additional internal and external conditions that may have an impact on your subject matter.

For instance, let's say we are looking at the future of autonomous cars in 2030. We might be compelled to only look up trends in the automotive industry. That is certainly a good start, but using STEEP allows you to look at patterns across other spectrums: what is driving people to buy into autonomous cars? What technology is available? What is the economic climate like? And how might those conditions drive policy changes? Understanding a policy environment could tell you whether certain ideas truly will be able to exist. STEEP essentially allows you to:

- Understand influencing conditions of today
- Understand influencing conditions in the future
- Categorize research and trends
- Categorize implications
- Provide an understanding of various aspects of an environment

Additional lenses and examples for STEEP include:

Social (and Cultural)
What people are saying or doing, their values and belief systems, viral topics, societal or group behaviors:

- Social movements
- Social media topics
- Social technologies

Technology
Advancements and progress in science, technology, and innovation:

- Emerging technology
- Information technologies
- Energy systems
- Medical devices
- Medicine
- Transportation

- Production techniques
- Patents filed

Environment (nature)

Referring to the natural world, resources, natural disasters, or ecological systems:

- Water table levels
- Air pollution
- Carbon indexes
- Impact measurements
- Climate change
- Extractive resource availability (petroleum, natural gas, minerals)

Environment (Context)

Can refer to a spatial context such as a kitchen, a family room, a building, or transportation:

- A living area
- Public space
- Inside a vehicle

Economic

Commercial activity, monetarily or financially driven systems:

- Interest rates
- Investments
- Employment rates
- Tax rates
- Stock market
- Consumer confidence indicators

Politics

Government viewpoints and structures of political parties, movements, or politics within an organization:

- Political parties in power
- Voting rates

- Shifts in power
- Type and number of political parties
- Political rhetoric
- Laws
- Treaties and alliances
- Business politics

Policy/Governance

Government, industry, or business laws and regulations that govern services, operations, or infrastructures; can also incorporate specifically legal topics:

- Regulations
- Terms and conditions
- Regulations under proposal
- Lobbying efforts

Values (and Ethics)

Attitudes and preferences toward work, life, and religious beliefs that influence individual or societal interests:

- Moral codes
- Religion
- Shared interests
- Political stance

Legal

Laws and government regulations (federal, city, state); can be referred to within policy or governance:

- Policies
- Regulations at the country, state, or city level
- Lobbying efforts

STEEP helps uncover additional conditions, influences, and interdependencies and allows you to dig into other aspects that could be critical to your strategy. When using STEEP, make sure to define what your time horizon is. STEEP can create very useful constraints or categories to help you to organize your thoughts and research as you traverse through time. It also gives you grounding and alignment on what factors are critical in building your future world. Later we'll look at how STEEP can be used in other foresight methods, such as the Futures Wheel and scenarios.

TRENDS, DRIVERS, AND SIGNALS

Designers are like anthropologists, which makes them well equipped for futures work. Through interviews and contextual information about systems, behaviors, and histories, we can look beyond the current manifestation of a problem and try to identify the root causes so that we know we are solving the right problem upstream. If we solved only for the current need or problem, we might overlook the drivers that created it in the first place, leading to potential recurrences of the same problem elsewhere in a system. Keep this in mind as you think about trends and how they've manifested in the world. They aren't necessarily design problems (though some trends definitely are), but they exist because something drove them to exist. And that driver created a small event that accumulated into a larger one over time.

I've mentioned trends and patterns a lot already, but it's important to understand the anatomy of a trend. The word *trend* may be more commonly associated with the fashion industry or with social media trends. But trends exist in many industries and locales and can be associated with economic shifts, politics, technology, or even societal values. The more repetition there is, the more impact it has. A trend is the culmination of many events accumulating until they can be recognized and labeled as a pattern. Trends are made up of signals, and drivers create signals. Since these terms can be confusing as far as which comes first, here are a few additional ways to think about how trends are produced, according to Institute for the Future:

- Drivers are to signals as clouds are to raindrops. Clouds, like drivers, overlap and converge and precipitate raindrops/signals. Those signals eventually accumulate, creating a trend (the storm).

- Drivers are to signals as diseases are to symptoms. Just as a symptom is a specific, observable result of disease, signals result from drivers and, when observed, call our attention to the presence of those drivers and how they might shape the future.

How far back do you go when you are trying to find the root cause (driver) of a trend? Well, that will depend on what you are investigating. If you are trying to adapt or innovate around a certain trend such as generative AI tools, you could probably just build a similar tool today, but understanding what actually drove this idea to fruition and why it continues to scale could be more useful. Creating another tool puts you into a crowded market (jumping on the bandwagon), meaning your differentiator will have to continuously outperform the competition. But going back to the principles that drove mass adoption could allow you to tap into what is making it work. Speed, large language models (LLMs) (*https://oreil.ly/viNe6*), accessibility, and fidelity of output were key drivers in the adoption of generative AI tools today. But other advancements in AI tools were also developing at the same time. Before AI became popular commercially, it was known as *machine learning (https://oreil.ly/NZZc2)*, which was just another term for smart logic algorithms that could perform complex operations quickly or generate predictions given a certain input. It was essentially AI, but in its most infant state. Machine learning was used in many ways as a prognostic or predictive tool to deliver options for how a system might react. In the mid-2010s, General Electric was using machine learning to develop some of the first digital twins[3] of jet engines and power turbines in order to understand how parts could fail when exposed to certain weather conditions. It proved to be faster and cheaper to emulate these scenarios inside a virtual machine rather than in a physical test bed. Today digital twins are used in many industries, from twins of cities to twins of humans, to predict behaviors. This early experimentation with analyzing multiple sets of data to generate answers laid the foundation for what is being used today in chatbots and other AI tools that accept natural language inputs as queries to get answers.

Returning to the boat analogy: imagine that our boat is starting to flood. We see water rising on the deck and think we can solve the problem by just scooping the water out with a bucket. But if we really wanted to stop the leak, we would seek out its cause. Perhaps we see a hole in the upper deck. But if we go to the lower deck, we might see another hole. By continuing our investigation to find the root cause, we solve the problem at its source. So let's say the hole in the bottom deck is the driver; the water droplets that began to form on the deck above, filling up the floor of the boat, are signals; and the water accumulating on

---

3 An article on the GE website (*https://oreil.ly/k8Vuk*) defines a digital twin as "a software representation of a physical asset, system, or process designed to detect, prevent, predict, and optimize through real-time analytics to deliver business value."

the deck above is the trend. You can see how one change leads to many small events that eventually accumulate into a larger event (the boat flooding). That's similar to how drivers, signals, and trends work (Figure 7-9).

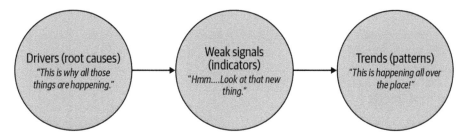

Figure 7-9. Drivers, signals, and trends

In Strategic Foresight, you might hear the term *weak signal*; it's what futurists seek to uncover early emerging patterns. A weak signal is similar to the first raindrop of an eventual rainstorm—it's something we see in the world that has the potential to become a larger occurrence. If you see one or two autonomous cars driving around your neighborhood, that's a weak signal that more may be coming. It's a sign that a company is testing the cars in your area. It's also a sign that your neighborhood might soon be crawling with robotaxis. I say *might* because a weak signal is as it states—*weak*. It has no strength yet because it occurs only in small instances (for now), but if you study the patterns around it (social, economic, technological, or political drivers), there is a certain level of potential for it to change and become stronger and eventually turn into a trend.

When we are thinking about the future, these are patterns we should be able to rely on because there is enough evidence (quantitative and qualitative) to give them credibility to be listened to. We can paint a much clearer belief system about a future world if we can show evidence around the accumulation of observations we are documenting. Weak signals may exist only in pockets, so they might be new and require more speculation to take seriously. They could dissipate and never be seen again, or they could transform into stronger patterns that become a trend that drives an entire industry. But what about *strong* signals? Strong signals have already accumulated enough momentum that they are trending or permeating cultures or geographies. Due to Tesla, there was a rise in ownership of electric vehicles all over the world. Once we saw sales increasing beyond the US, the rise in ownership could definitely be considered a mega trend. However, even as it scaled domestically outside of California, it was already surpassing macro into mega. It's much easier to identify mega trends that everyone is talking about and competing in than the ones that are hidden or emerging. Futures shouldn't ignore current evidential patterns because those become drivers toward other trends as well. We just tend to seek out the weak signals due to the desire to uncover unseen or new threats and opportunities. In time, with enough practice, you will learn to identify these patterns, causes, and shifts, internally categorize them, and become faster at seeing these indicators as they emerge.

Types of trends

You might already have an idea of what a trend is, as you see trends play out on social media or in fashion or technology consumerism, but there are different types of trends that can be categorized by their size, influence, intensity, and reach. There are mega trends, macro trends, micro trends, and fads (Table 7-1). Each type of trend has a different set of characteristics that can be used to describe it based on size or impact.

Mega trends are the largest type of trend, and you've certainly already heard of them because they've traversed national borders, media, television, magazines, and articles. Everyone is talking about mega trends, and they can impact entire industries for decades to come. An example of a mega trend might be consumers buying electric cars in several countries. Teslas are now seen all around the world. And other car companies have now started building electric cars due to this mega trend. In the automotive industry, the current mega trends are called ACES (automation, connectivity, electrification of the drivetrain, and shared mobility); these four mega trends are driving every transportation manufacturer in the world and will continue to drive them for the next 20–30 years.

Macro trends are a bit smaller. Sometimes they can be defined by regional or national borders, but they are still growing; maybe they just haven't reached a global or industry-impacting level. A trend in the US might not be a trend in China, so it could be considered macro because it is confined to only a small part of the world. It could involve several nations but not yet be global. An example of a macro trend might be the veganization of recipes for traditional meals.

Veganism has been rising in popularity for decades and is becoming increasingly mainstream. And more and more we are seeing vegan recipes for familiar foods (e.g., vegan macaroni and cheese) across social media and food blogs. But in addition to alternate recipes, we're also seeing vegan versions of meals packaged in grocery stores. You might find these in parts of the US or Europe but not in Africa or Latin America (though that's quickly changing). As of today, the veganizing of many traditional foods could be a macro trend that is just about to become a mega trend, and at some point it will be.

Micro trends are even smaller and possibly haven't existed very long. Before Facebook was a social media giant, there were micro trends of social media platforms that grew in the US, such as Friendster and MySpace. Social media as a communication platform was merely a micro trend but quickly grew to a macro and mega trend as more people started using it and creating competing platforms. An example of a micro trend might be paying for physical products or services with Bitcoin. Not many brick-and-mortar stores take this form of payment, but there certainly are some convenience stores and gaming outlets that will accept it. Because of its extremely limited use in physical retail, it's a micro trend, but there are enough instances of it in the world to say it is a trend and that more stores might start accepting cryptocurrencies at some point.

Fads are brief and transient—they come and go very quickly. Some turn into micro trends and grow to become mature trends. But we typically use the word *fad* to describe social and fashion trends that appear and disappear quickly. You might even say a fad is a weak signal. Fads are not to be ignored but rather should be monitored because, just like weak signals, they can evolve into micro, macro, and mega trends. They might not be influential yet, but they could become the root driver for something much more influential later. An example of a fad might be hoverboards. Hoverboards grew rapidly in popularity and may have looked like a mega trend at one point because they were seen in many countries, but they didn't last long. Due to problems with their batteries exploding on airplanes, safety issues, and their being banned in certain cities, the hoverboard fad faded out very quickly.

Table 7-1. Mega trends, macro trends, micro trends, and fads

	Fads	Micro trends	Macro trends	Mega trends
Description	Short-lived and quickly appearing and disappearing trends	First signals of emerging patterns; may be concentrated in groups of networks or within a neighborhood, city, or region	Beginning of compounding trends appearing across a society with scaling momentum	Major cultural, societal, economic, political, or technological change affecting an entire industry or many countries' populations. Traverses several national, regional, and digital borders.
Duration	1 month–2 years	2–5 years	5–10 years	10–30 years
Impact	Peer groups, small networks	Localized within small populations or within a city or a small region	Country or large regions (EU), multi-country or multi-platform. Due to its scale, has potential to evolve into a mega trend.	Global or internet-wide impact. Shifts industries, opinions, and views
Example	Diets, clothing, hairstyles, fashion	City- or country-mandated procedures, slang, artisan foods or establishments	Use of VR/AR only in certain regions or communities, stores that accept Bitcoin	Climate change initiatives, artificial intelligence, TikTok

The future is a moving target. It's constantly changing, and there are always uncertainties and surprises. But we can at least develop an informed opinion about the future through the data we gather. Once we have synthesized and prioritized that data, we can begin building out potential future worlds and can start to assign characteristics, rules, conditions, and feelings about those worlds. And while each of those worlds may have some grounding in information we gather today, we must also recognize that trends are agnostic of a good or bad scenario and do not label themselves as a threat or an opportunity. They are merely what we are observing. It's how we frame trends within scenarios (which we'll discuss in the next section) that gives them a particular disposition.

For example, there is a trend of Gen Zers using TikTok to search for subjects instead of using Google Search. We don't know whether this is a good or bad trend or what kind of world it lives in. Is it part of a dystopia? And whose dystopia is it? Google's? Our society's? Or is it a transformational and innovative scenario in which this new platform becomes the next global search engine? You might have a strong opinion about what this means, but it could be the inverse of what others think. Any trend can be positioned in relation to a particular theme, depending on whose world it is and whether you perceive the trend as beneficial or detrimental. This trend could play out into a positive scenario for industries changing the landscape of search and social media and open up new opportunities for tech companies and advertisers. Alternatively, it could become the death of the search engine and collapse the validity of information people are accessing, leading to global strife and destitution. Either scenario is valid and possible and could be paralyzing if you play it out too deeply. That's why it's important to consider many potential outcomes and then make informed decisions about how to deal with them today and apply guardrails or policies to steer them in the direction we prefer.

If or when you need to introduce the concept of trends, drivers, and signals to a new group, think about *how* you will introduce it and how much you need to expose to the group during this phase. Some people will just want to know what the trends are, so you might not explain how drivers and signals influence trends. They may not need or want to know these nuances. You could just show them the synthesized trend cards, include significant data points or quotes, and list the sources (Figure 7-10). Sometimes that is enough to build the trust and belief system that the trend is real and worthy of paying attention to. You can do the deeper analysis behind the scenes but only showcase the end results for simplicity and to abstract the analysis into a language they can understand. For those who are more familiar with trend research, you can introduce the deeper anatomy and drivers of what formed the trend, which can open up conversations about other ways to innovate. This isn't to say you can't innovate without exposing the roots of trends; it is just a matter of what your audience needs to understand at that point in the process to align and move forward. But naturally you should always have evidence that supports a trend.

Figure 7-10. Trend cards (source: Futures Platform (https://oreil.ly/U13uQ))

Quantitative versus qualitative intelligence

During your research process, you will start to home in on various types of futures intelligence and data. To supplement your inquiry, you may also want to include some validation of the information you've found. Quantitative data attached to your trends can enrich and describe the gravity or breadth of a particular trend in the market, thereby giving more credibility to the information as you discuss it with your audience. These quantified markers can also prove useful in the prioritization stage when you are trying to decide which trends have more weight or probability of scaling within a particular time horizon. Some examples of quantitative data are:

- Number of startups that exist around this topic or technology
- Size of market investment (how many millions or billions of dollars have been invested in a topic, trend, or technology)
- Years of growth (of a company, product, or technology)
- Scale and reach of growth (how many places are involved in this trend or are using this technology)
- Number of instances of the topic (across different sources)
- Number of academic whitepapers
- Number of experts who support or agree on the topic
- Number of sentiments on social media
- Number of patents filed

There is no standard minimum or maximum number for how many sources or experts you speak with. That will be up to you to decide based on the time you have to research and the evidence that you or your organization needs to make the trend credible (just two sources that say something is trending may not be enough, but $2 million in market investment may be impactful). But do take into account researching a variety of resources to validate that something is in fact being discussed widely, as well as looking at who is talking about it and in what regions. If a pattern exists only within your city and nowhere else, then it is probably not a mega trend (yet). Similarly, if several different sources, experts, and investments are attached to a technology but it is owned by one company, the variety of instances can validate that the technology is significant enough to be labeled as an "emerging" trend.

Each figure you present can express the scale of investment, growth, and impact of the topic over time. This will automatically assign a weight or value so your team can perceive how important, realistic, or threatening the pattern is. This can be helpful both in validating the size of a trend and when prioritizing trends in later stages. For example, you might not think that a trend of edible insects is going to be very big in the US, but if you saw that there was $13 billion in investment, 20 noted food experts are saying it is going to explode in the next 5 years, and the same opinion has showed up across 30 trusted news sources, you might change your opinion. The greater context within which the trend is scaling, and whether it's relevant or impactful to your particular business or industry, is also important.

Qualitative data is just as important. This is information you are gathering from your own secondary research as well as from interviews of experts or everyday people (primary research).[4] Examples of qualitative data include:

- Quotes from domain experts
- Interviews with users or customers
- Sentiments on social media
- Customer feedback from customer support records
- Opinions of news anchors, scientists, and academics
- Product reviews
- Quotes from articles
- Historical documentation of events (articles, videos)
- Trend reports (which typically include synthesized quantitative and qualitative data anyway)

Developing a vocabulary of opinion and numbers can only enhance the strength and exposure of what a trend is composed of. The more you have of each type of data, the stronger the case for this trend being at the scale it is and for its potential to continue scaling up or down.

Together, quantitative and qualitative data can really bring trends to life. Unless you are working with a mega trend, you might need additional information to validate or enhance a trend that can be seen only in small quantities (weak signals) across sources.

Prioritizing Trends and Other Futures Intelligence

Once you've gathered all your trends and futures intelligence, you might find yourself with heaps of information. You might have global mega trends as well as macro trends across different countries, interviews, market data, trend cards,

---

4 The Qualtrics website (*https://oreil.ly/VgRjV*) defines secondary research, or desk research, as "a research method that involves compiling existing data sourced from a variety of channels." The University of Southampton Library website offers this definition (*https://oreil.ly/tY6z-*) of primary research: "Primary research involves gathering data that has not been collected before. Methods to collect it can include interviews, surveys, observations or any type of research that you go out and collect yourself."

reports, and even domain expert opinions. How do you make sense of all this information without going crazy and overwhelming yourself and your team? There are many ways to prioritize information, including holding meetings and workshops with your stakeholders and clients to decide which information is relevant. A prioritization method or matrix is also very useful, and I recommend that throughout the entire futures process you align on how you will make prioritization decisions, who will need to be involved, and what frameworks or data you will need to make those decisions collectively.

PROBABILITY VERSUS IMPACT MATRIX

A *prioritization matrix* is a grid that allows you to plot your information across two or more axes using specific criteria. One type of prioritization matrix you can use is the *Probability Versus Impact Matrix* (Figure 7-11). Before you start, it's important to define each axis as specifically as possible. The y-axis is typically *Impact*, and the x-axis is typically *Probability* or *Likelihood*.

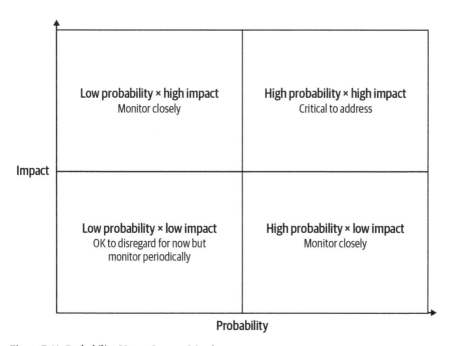

Figure 7-11. Probability Versus Impact Matrix

Try to define the kind of impact you want to measure:

Impact on the business
> What is the likelihood this trend will impact the business's revenue stream, bottom line, existence, or operation?

Impact on the product
> What is the likelihood this trend will impact the product's success, its user experience, or its rollout strategy?

Impact on the user
> What is the likelihood this trend will impact our customers, users, or citizens? Will it harm them or protect them? Will it deteriorate the user base?

Impact on other technologies
> What is the impact of this trend on other technologies or resources (e.g., the impact of AI on other non-AI-enabled machines)?

Similarly, try to define the type of probability you want to measure:

- What is the probability or likelihood that this trend will scale to a $3 trillion market investment?
- What is the probability that this trend will become an international mega trend?
- What is the probability that this trend will not scale because there isn't a cultural interest in it?
- What is the likelihood that this trend will triple in active user engagement?
- What is the probability that this trend will still be a mega trend in 10 years? Or that this fad will become a macro trend in 2 years?
- Which trends will be likely to scale only in our city?

Being clear about the parameters you are using to prioritize will place you in that world so you can decide how to prioritize given the conditions of the matrix. You can also divide the matrix into as many quadrants as you want. You can have a simple 2×2 matrix (this is the easiest) or a 3×3 matrix with six quadrants. If it helps to have a 3×3 matrix (with the categories High, Medium, and Low, as in Figure 7-11), then use that. The more quadrants you have, the greater definition you can have around how to prioritize.

Other variations of criteria could be based on what the evaluation priorities are in the project. Maybe they have to be very specific around a regulatory environment, user context, or economic impact. Impact is great because it measures a type of change as an effect of a trend, pattern, or technology, but it can be described in different ways as well. Other variations could be:

- Value versus impact
- Regulatory readiness versus customer need
- Popularity versus environmental impact
- Number of users versus number of startups (using a technology)

Fundamentally, the matrix is a way to isolate a particular set of data based on the data's existence and value in the future world. It allows you to reduce your body of research so you can focus on your focal issue and on all the important conditions and influences that could impact or shape the future. It's flexible and modifiable, however you need to use it to get the job done. And you can use it at any point in the process to help you prioritize any kind of data (research, scenarios, ideas).

Some trends and data points can have an automatic weight or value due to a value you assign to them. Let's say that "the rise of quantum computing" is a trend you have on your list. By also including how much has been invested, how many startups are working on it, and how many distinguished articles discuss the rise of quantum in the next decade, you can assign a particular value that shows just how fast this trend is scaling and how valuable it is today. If that value is high, and if the numbers say that there has been a steady level of investment over the last decade and that that investment is accelerating, we can speculate that this trend will continue to scale over the next 5 to 10 years given its current trajectory. We now have a given value that will allow us to say that this trend is real and getting bigger and that we know the probability of it existing in 2030 is high. Now we can just decide how much impact it will have on something we define (our business, our customers, other technologies, etc.).

Using weighted values with your trends can take the subjectivity out of the analysis when you're trying to decide a trend's importance in the future. Of course, there will always be some unknowns, especially with micro trends and other weak signals. We may lack a lot of quantitative data about a trend and thus may have to speculate or triangulate how other trends and data might enable it to scale into a macro and mega trend. But that is OK! There are many assumptions we have to make about the future and the information we know today. Again, we do this every day. We calculate some information that we can depend on in our head, and we make educated guesses about other information. But the accumulated information allows us to make better judgments about the future and to at least consider the impact or consequences of other events that may occur.

The caveat to using this framework is that you must be clear about how you are measuring the value of the information on the grid. Is the weighting system a subjective opinion, or is it being quantified by a particular formula or algorithm? Be clear about that so there is no confusion as to why something is being prioritized.

Remember that the intent of the matrix is to analyze which trends have the propensity to expand or contract in your future world. There may be some H1 horizon trends that are due to die out and some H3 horizon trends that are sure to expand. The matrix is also a useful way to quickly focus on what information you want to bring into the future world and what you want to omit. However, just because you omit something from a high probability and impact score doesn't mean it's irrelevant! It just means you are choosing not to focus on it for the purposes of your project's focal issue right now. If you incorporated every single trend you think will exist in the future world, you could easily get distracted, overwhelmed, and confused about which one to pay attention to, and you might never come to any strategic conclusions.

There are several benefits to using a prioritization matrix:

- A prioritization matrix is useful for down-selecting and prioritizing data.
- It can provide more rigor in how to calculate priorities.
- It narrows your research to the important topics you want to explore.
- It provides a structured and logical value measurement system.

There are also some cautions associated with using a prioritization matrix:

- Without quantitative data, it could seem subjective.

- Axes must be defined or else there could be confusion as to what *impact* or *likelihood* actually mean.

- It could feel like you are ignoring important research outside of your scope.

- People tend to want to put everything in the high probability/high impact corner, so you'll need to guide them to make sure they don't over-index on everything.

Another way to prioritize your data is through simple voting mechanisms. This approach may feel nonscientific or subjective, but not every methodology is rooted in a mathematical principle or logic. Speculation is inherently a guess, and we make subjective decisions all the time. Of course, we try to use evidence to back our dispositions, but sometimes it's necessary to bring in a subjective argument for or against something to move the ball forward. A CEO might be interested in particular kinds of trends or technologies, perhaps because of financial or technical feasibility or resources. They might not care about trends outside of their industry at all. This is OK, though it would be good to note that trends exist everywhere and can influence any industry in a number of ways (Figure 7-12). If you were a gaming company and looked only at gaming trends, you might be blindsided by trends happening in entertainment, fashion, or other nondigital gaming communities that make their way into the digital gaming realm. This is the case with gaming platforms such as Roblox (*https://oreil.ly/ TCDBa*) and Fortnite (*https://oreil.ly/cXUvN*), as brands and artists infiltrate these virtual platforms to launch virtual stores (*https://oreil.ly/2izGr*) (Figure 7-13) and concerts. Scanning across many industries and picking up on signals, especially those launched by major brands or influencers, can reveal the signs of adjacent or ancillary industry trends that will be coming into your domain soon.

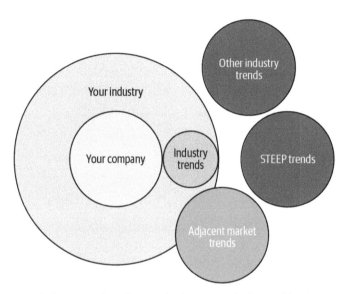

Figure 7-12. Intra-industry trends and external adjacent trends that could influence a company's future

Figure 7-13. In May 2022, Gap launched a virtual Gap Teen store on Club Roblox, a community site for the Roblox game

Trends, drivers, signals, and other futures intelligence make up your body of research that can be used to begin building future worlds. Giving yourself enough time to gather and collaboratively analyze this information will be the lifeblood for making futures work in later stages of the process. Chances are that as you research and discuss these patterns, you will want to start brainstorming ideas quickly. Try to hold back from jumping to any solutions too early. Solutioning too early could mean short-sighted thinking, and you might overlook important scenarios for how those patterns actually could play out in the future. If you truly want to innovate, you'll want to try different ways of exploring, processing, and digesting the information you have. You'll want to test ideas, consider implications and various scenarios, and utilize additional tools and frameworks to help you have a more holistically informed vision.

Insight Versus Foresight

When thinking about the research necessary for futures work, it's important to make the distinction between insight and foresight. Typical design or marketing research teams are traditionally focused on insights gathered from user interviews and current market trends; this is data that is more anchored in real-time trends and maybe looks out a few months or a year or two. Foresight work, however, traditionally tends to look further out, using a variety of sources to piece together a picture of distant future landscapes. Foresight teams also traditionally might operate separately from insight teams and are used for long-term strategic planning. Insight teams are used more for immediate deliverables. However, both types of teams can benefit by sharing data and capturing snapshots of the current world so that they can look beyond the current horizon.

—Sarah Owen, Global Futures Advisor at Soon Future Studies and leader of a global community called Futures Friends

Sarah Owen has a variety of approaches for helping organizations interpret trends and signals and bridge the gap between insight and foresight teams and data. One of these is the creation of a private Instagram account that the client can log in to and post trends and signals. The account acts as a safe, closed-wall garden through which everyone in the company can interact daily, communicating and commenting as posts appear in their feed. The use of the popular social media platform allows the teams (and organization) to be plugged in and walk the journey as futures projects progress, which leads to greater collective buy-in and greater understanding of context and approach, and ultimately to more trust in the process and outcomes. In an interview I had with Sarah, she shared this about her process:

> I like to make sure I'm clear about if a project is insight-based and going into the hands of people that work on insight teams, or if it's more futures focused and going into a foresight team, because if you're trying to learn insights in a foresight team, there's going to be a disconnect. Some want clear implications; they want a roadmap. And if you're introducing foresight into a team whose KPIs are anchored on the next 6 months or 12 months turnaround, you're already lining up for disaster.

As noted in Chapter 6, having a repository for your research is critical.[5] Whether it's an Instagram feed,[6] an Excel database, or a Notion page, having a place you can tap into for analysis and synthesis is important. But it takes rigor and constant awareness of your boundary conditions to ensure that nobody gets overwhelmed and to avoid the occurrence of "scope creep." This process shouldn't be too dissimilar from traditional design research, where you want to keep the problem statement, project goals, and user needs in mind and recognize when it's time to stop researching, synthesizing, or designing.

In my interview with Sarah, she discussed how we use boundaries as vehicles to enable open exploration and as constraints to converge thinking:

---

5 A repository is a place for you to store your trends and data. It can be a folder, an Excel spreadsheet, or some kind of database with tags to categorize each piece of information (source, author, date, topic, type of trend, signal or driver, number of instances). You can devise your own system of categorization, but the key is to have someplace to keep everything neatly so that you can access it later as evidence or reference for trend cards, reports, or decisions that were made toward the future scenarios or vision.

6 An Instagram feed is just one type of social media platform that could be customized to act as a repository of articles and information posts. It can be a living document of signals or trends that you can scroll through. Using a platform like this allows you to collect stories and posts from many sources so that you can scroll through them over time.

Without anchors or really clear boundary conditions, you can end up in a rabbit hole of research for weeks. But you don't want to add on too many boundary conditions straightaway because you're cutting yourself off from looking at the world objectively. So there's an art to it. As you narrow down and synthesize within the trend analysis stage (which requires a second check-in with the client to reestablish the boundary conditions), you need to make sure you are still going in the right direction, or if it requires pivots, because sometimes new research emerges and you have to follow that trail. Reclarifying the boundary conditions after the client has seen and had access to the expensive research is really valuable. Also, within the trend analysis, we have a data bank that we use where we tag and have a taxonomy of tags that we use for signals, so we're always collecting those, and we can go in and click on "Gen Z" or "sustainability" or "South Korea" or "Metaverse," and that helps to aggregate the latest and the greatest information. There are a lot of platforms out there that do the same thing that are automated, but we still cherish the art of doing this manually, because we are using different platforms, not just the traditional web. There are so many signals we grabbed from Instagram and TikTok that speak to true human behaviors that you can't get from a New York Times *article or the* Verge*. We are big advocates of getting out of just doing desk research and getting signals from the real world. Which I think we forget is possible.*

There are many approaches to organizing, analyzing, and synthesizing trends and other futures intelligence, but in the most successful cases, organizations will learn a lot along the way (and in turn become more futures literate) regardless of the methods, terminologies, or principles you use or expose them to. Exposure to the research in a way that they can easily relate to and be inspired by can truly transform the way the organization operates, thinks about futures, and ultimately develops its own stance as it moves through the process. Try to make sure you have a clear approach for organizing and prioritizing, and use it as often as needed to help bring clarity to the information you are synthesizing.

Once you have prioritized, segmented, and organized your data, you have a basic set of information that you can use for scenario planning and world building. Remember that this information is agnostic of a theme. It's neither dystopian nor utopian in nature. Only after you have assigned the information to a specific future scenario or disposition (which we'll discuss in the next chapter) can it be perceived as positive or negative. Once you have isolated the set of information that you think will be prevalent in your future time horizon, don't throw the other data away. The prioritization task is only so that you can have a set of information that you can work with for world building. That doesn't mean it's final and that it is the only information you need moving forward, but prioritization allows you to create a digestible set of information to work with. You may want to revisit the other trends and signals later, especially if things change as time goes on. A weak signal could quickly become a strong one in a short amount of time and become a major influencer in your future time horizon. And a signal that you originally scoped as a powerful influencer could peak and decline in the next year, removing it from your selected set of trends you care about. The future is dynamic and always changing, but we need to try and make sense of it with the information we have, so keep an open mind as you prepare yourself for the scenario planning phase.

In Closing

Horizon scanning and trend analysis form the core of where research begins and operates within Strategic Foresight. While some of the tools, data, and horizons may differ from how you might traditionally perform research today, the mental model is still very similar in that we are parsing information about the present to understand what might be needed in the future. Be careful not to get too overwhelmed by the amount of data you capture, as this process could theoretically be boundless if you are not setting the right priorities and constraints. To truly be futures-ready, you will want to make trend hunting and scanning a daily practice, one that becomes an act of nutrition for you and your organization. In the next chapter, we'll discuss how to use this information to start developing future scenarios.

World Building with Scenarios

Everyone naturally thinks in scenarios. With every goal that we plan and every decision we make, we are subconsciously playing out scenarios in our head so that we can be prepared. We do this by asking "What if?" questions: *What if this happens? How will I respond? What if this doesn't happen? How will I respond then?* Throughout our lifetime we're able to quickly visualize scenarios in our head and make decisions about which are most likely to occur, which we might want to be prepared for, and which can drive different outcomes. And as those scenarios present themselves in reality, we are able to face them either with a conscious plan or with some understanding of what to do next. Without this internal scenario planning, we might end up being surprised by and unprepared for certain events and have to make decisions on the spot. Sometimes that works out well for us, and at other times it does not. The more scenarios we consider, the less surprised we might be in the future. In terms of Strategic Foresight, scenarios are merely visualizations of stories that could play out in the future. They have themes, dependencies, emotions, constraints, and many other facets. We visualize them in words or images so that we can investigate what we might do if or when they arrive.

Let's say we are driving to a state park in a rental car, and there are long stretches of highway with few gas stations. Our gas is running low, but we think we can make it to the park without having to stop at the next gas station. We want to arrive early at the park so that we can make the most of our day there. In our mind we might visualize a few scenarios. One scenario might be that we make it to the park with no problem. Our gas level is fine, and we can enjoy the day as planned. Another scenario is that we don't actually know this rental car that well, and it might burn gas faster than we think, meaning we could run out of gas before we reach the park. The last thing we want is to be stranded. Our brain starts to do several calculations in an attempt to assess all the risks and opportunities, even going so far as to play out the nonpreferable scenario of being stranded. If we were stranded, we do have a cell phone that we could use to call for help, but would we have a signal? We then try to balance other factors, such as travel time, maximum time of enjoyment at the park, and the conditions necessary to return home afterward. And we are doing all of this while concentrating on the road ahead. With each second that goes by, the conditions change, our gas slowly depletes, our options change, and our recovery time lessens.

Scenario planning is no different. As we move into the future, conditions are changing every minute. Each of those conditions continues to affect our decisions and our outcomes. We have to constantly manage perspectives of the future, understand multiple dependencies, and play out alternative scenes based on what we know about the road ahead. Some of our knowledge is unknown and speculative, and some of it is backed by evidence or memories that we have confidence in (our driving ability, our knowledge of the route, our knowledge of the vehicle we're driving). Some of our knowledge is hearsay—maybe we heard the rental car agent say that this vehicle has great gas mileage. Since the agent is probably a subject matter expert (an SME) on that issue, we might take that statement as fact and use it to build our own confidence about how far we can go. While we are processing all this information to inform our decisions while driving, we are constantly trying to manage our risks. But at every point along the way, we have to reassess our situation and revisualize the scenarios based on where we are on our journey. If we're smart and thorough, the nonpreferable futures might not occur, and we will be able to make it safely to our destination. But if a nonpreferable future *does* occur, hopefully we have a plan to address it and will still reach our destination safely.

Scenarios are built from the research we've gathered, synthesized, prioritized, and projected into a future time horizon. They become the basis for the conditions, rules, dependencies, and behaviors that could exist in a future world. But as we mentioned in Chapter 7, trends are agnostic of a theme or polarity. They do not call themselves positive or negative. Thus, when we put a trend

in the future, depending on who is looking at it and how they are perceiving it, the trend can be good or bad. The first step is to understand what those trends are and how they might influence each other in the future ecosystem. Scenarios help us inject a polarity or set of conditions that sway those trends in one direction or the other. A trend of buying local vegan organic food is a consumer trend. It doesn't tell itself it's bad or good for society. From a health perspective, it seems like a positive trend transforming our diets and easing the pressure on farmed and slaughtered animals, thereby reducing the impact of industrial animal-processing waste disposal on the atmosphere. On the other hand, a surplus of local vegan food production can add pressure on supply chains; as demand increases, the industrial measures used to meet the rising demand make this trend potentially threatening to competing producers, add more materials and waste to the supply chain, and could be just as detrimental to the environment as the production of animal foods. But this consumer trend is just the *trend*. It's the scenario that we build around the trend that can shift it into realizing its positive or negative implications in the future. This may at first make you dizzy once you start to realize that any trend can have any polarity or go in any direction. But this is also a symptom of the *weight of alternatives*, which we have to continue to manage as we explore alternative futures, scenarios, and implications. As we develop scenarios or any other generative output, we have to keep in mind that we have ways of prioritizing what's most relevant or most important to consider for the purpose of the work we are doing.

Scenario Formats

Scenarios can be illustrated as different formats. Consider which format will be the most impactful for your audience. Even if you and your team have been talking about the future for some time and using illustrations or images, much of that visualization is still happening inside the audience's heads. Choosing the right format can make all the difference in immersing people into that future world and allowing them to look around, empathize with the people, plants, and animals that live there, and make better decisions about what they want to solve for. A few common scenario formats are narratives, illustrated images and scenes, and immersive and interactive experiences.

NARRATIVES

One of the more popular formats to describe a future scenario is a short narrative. Depending on who your audience is, how you are publishing or presenting the information, and what's important to highlight, narratives can be any length. Some are expressed as a day in the life of a person or system in the future. Some are more general descriptions of the environment and the impacts that have occurred over time. Some are written as a historical account detailing what happened when certain events or trends changed the world. If you've conducted a good diagnostic and spent more time with your audience, team, or client, you might have some clues about what format works best for them. Maybe the presentation is better told through a video, or maybe it needs to have specific information highlighted, or maybe it's just a short paragraph. If you're unsure about the required or desired format, length, or details, you can always ask or negotiate what would be the most impactful way to illustrate the scenario.

In 2019, Arup, a sustainable development futures consultancy, released a report titled *2050 Scenarios: Four Plausible Futures* (Figure 8-1). It includes a series of four divergent futures—Humans Inc., Extinction Express, Greentocracy, and Post Anthropocene—that range from the collapse of our society and natural systems to humanity and nature living in sustainable harmony. The Greentocracy scenario is described thusly:

> *Climate action and biodiversity recuperation are the top-line of every national and transnational agenda. The results of the galvanised global efforts have been unprecedented for the environment, but not without significant sacrifice from people who are realising the trade-offs did not quite work out for them. Humanity now lives in self-imposed servitude to the environment under the mantra of "happy planet, happy people."*

GREENTOCRACY

Driven by extreme ecological regulation, societies are now highly divided, regulated and unequal, with most citizens increasingly disillusioned; they are mere pawns in the Greentocracy. Since the first civil protests in 2040, the SunGrown movement has grown significantly, demanding access to the natural, non-LABFood enjoyed by the affluent part of the global population, as well as more access to nature for leisure activities. However, the requirement to expand farmland and roadways to achieve these demands would go against the principles of land-use regeneration and has therefore been vetoed by regulators.

As a way of limiting environmental impact, governments continue to discuss the controversial topic of global population control. Some legislation has already been tested, but ethical concerns prevent large-scale implementation. The results turned out to be inconsequential, as the poor state of society is already enough to prevent people from wanting to bring children into the world.

9.5bn
global population

high wealth gap

1.5°C
increase

medium global cooperation

stable weather

97%
clean energy

2050 SCENARIOS

Figure 8-1. The Greentocracy scenario from Arup's 2050 Scenarios *report*

Nordic Cities Beyond Digital Disruption, a report published in 2015 by the Smart Retro project and the think tank Demos Helsinki, discusses how Nordic cities have been impacted by digital services, urbanization, and the changing landscape of how people navigate, live, and work today and how that might change in the future. The report presents three scenarios that detail how the world could change over the next 25 years and why Nordic cities can achieve success amid political, social, or technological changes. In "Scenario 3: Crisis and Recovery" (Figure 8-2), the narrative begins by explaining the consequences of the events of the early 2020s and how the pandemic and its global impact could affect several behaviors, technologies, and investments in certain sectors such as energy, mobility, retail, policy, and public institutions. This section of the report includes a timeline of events from pre-2020 to beyond 2030 to illustrate the sequence of implications and events that could lead toward this scenario.

Narrative (text-based) scenarios can include supporting information that is needed to convey different dimensions, impacts, or details of the scenario. This additional data serves to support the story and can help your audience gain a holistic view of what that scenario looks and feels like. Much like a science fiction novel, a narrative scenario builds on the history, actors, and influences that have shaped this world. But scenarios are merely a window into the future. They can behave as vignettes, or they can be used to build a larger story or argument that you can strategize around.

Some benefits of using a narrative to describe a future scenario are:

- It allows you to create a story around a set of conditions.
- It allows you to integrate many aspects, including actors (future personas), policies, and environmental conditions (STEEP).
- It's cheap to facilitate and produce (it's just writing a story).

There are also some cautions to consider when using a narrative format:

- Depending on the length of the narrative, words without imagery can be a bit boring and may not achieve the impact you're hoping for.
- A narrative depends on the author and style of writing to make it sound and feel impactful.

SCENARIO 3: CRISIS AND RECOVERY

THE GLOBAL ENERGY CRISIS in the early 2020s pushes most countries into deep recession and provokes unforeseen levels of migration from many vulnerable regions towards wealthier countries. Public expenditures soar momentarily as a vast array of companies go bankrupt and the availability of basic commodities decreases in many countries. The increasing flood of climate refugees puts a serious strain on the economies of European countries, still recovering from the previous slump. Over the course of the following years nations are forced to radically change their governance structures, industrial policies and energy systems. The crisis acts as a catalyst for the breakthrough of many ripened solutions linked to energy efficiency and renewable energy. Pioneer cities and regions manage a faster return to a steady development path thanks to their investments into energy efficiency, renewable energy and sustainable urban structures. Yet recovery also demands a new notion of progress and an economic system that combines money and bartering. Resilient local structures gradually replace many systems that had operated within the context of nation states, extending from many public services to energy and trade.

Crisis → Behavior → Technology → Investments

ENERGY: Decentralised local production of renewables, with solar power and bioenergy at the core.

MOBILITY: Sharing rides, vehicles and transport of goods becomes an important part of mobility. The total amount of kilometers driven decreases significantly.

CONSUMPTION: A momentary but radical decrease in purchasing power changes consumer behavior. A part of the population shifts to a barter economy in basic goods, but slowly returns to a money-based economy.

RETAIL: Different forms of P2P commerce; big retail companies reform their logistic systems and focus on smaller neighborhood stores.

BUSINESS: The crisis forces a large number of companies to go bankrupt and the markets are partly consolidated into the hands of a few of the strongest actors.

New regional actors emerge to maintain energy and transport systems.

WORK: An increasing number of people receive their income from various simultaneous sources and make up for missing revenues by bartering.

POLICY AND PUBLIC INSTITUTIONS: Vast differences between regions and nations. In well-prepared regions, the role of local politics grows and citizens place higher trust in local than in national politics.

NORDIC SOCIETIES: The energy crisis forces countries to use public funds to subsidise critical functions such as energy production and logistics, which means that welfare structures are slashed. Some of the services are maintained by the voluntary sector. At the time of recovery (2030) basic income is introduced in order to secure workforce for the maintenance of locally produced public goods.

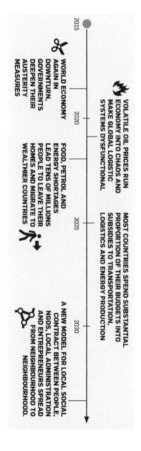

2015

WORLD ECONOMY AGAIN IN DOWNTURN. GOVERNMENTS DEEPEN THEIR AUSTERITY MEASURES

VOLATILE OIL PRICES RUN ECONOMY INTO CHAOS AND MAKE GLOBAL LOGISTIC SYSTEMS DYSFUNCTIONAL.

2020

FOOD, PETROL, AND ENERGY SHORTAGES LEAD TENS OF MILLIONS PEOPLE TO LEAVE THEIR HOMES AND MIGRATE TO WEALTHIER COUNTRIES

2025

MOST COUNTRIES SPEND SUBSTANTIAL PROPORTION OF THEIR BUDGETS INTO SUBSIDIES TO TRANSPORTATION, LOGISTICS AND ENERGY PRODUCTION

2030

A NEW MODEL FOR LOCAL SOCIAL CONTRACT BETWEEN PEOPLE, NGOS, LOCAL ADMINISTRATION AND ENTREPRENEURS SPREAD FROM NEIGHBOURHOOD TO NEIGHBOURHOOD.

Figure 8-2. "Scenario 3: Crisis and Recovery" from Nordic Cities Beyond Digital Disruption

ILLUSTRATED IMAGES AND SCENES

An image can be a nice complementary artifact to a narrative text-based story when describing a scenario. You can use hand-drawn illustrations or a series of images taken from the internet. There are also many generative AI art tools such as Midjourney that will allow you to create images based on prompts in mere seconds. How you stitch together those images should be based on what you know about the intent, the audience, and the impact you want them to have as an aid to your scenario.

In 2021, the Federal Transit Administration awarded a grant to the San Francisco Municipal Transportation Agency (SFMTA) to develop a National Transit Adaptation Strategy to support the public transportation industry's adaptation through and beyond the pandemic. The SFMTA partnered with Institute for the Future (IFTF) on the year-long project; as the institute's website explains (*https://oreil.ly/mFQfh*), "the work aimed to identify mobility needs and target market segments while developing messaging to rebuild confidence in public transportation and increase ridership. IFTF's additional goal was to explore what it means to build a future-ready public transportation agency that embeds foresight in its plans for rapidly changing rider needs, climate-related shocks, or drastically altered commute patterns resulting from new ways of working." One of the resulting deliverables, a report entitled *The Future of Transportation: Five Scenarios*, featured both detailed scenes of future scenarios (Figures 8-3, 8-4, and 8-5) and stories about each scenario that painted a vivid picture of how transportation could have responded or adapted to the COVID-19 pandemic.

Figure 8-3. "Scenario 1: Shifting to Market-Based Mobility Systems," from Institute for the Future's The Future of Transportation: Five Scenarios: *"Public transit in most US cities never recovered from the impact of COVID-19 and the widespread shift to remote work in the 2020s.... Focusing on profit turns attention away from money-losing routes, services, and investments. It shifts focus to high-margin services and sometimes draconian cost-cutting measures. Running a public transit system for profit leaves behind the poor and marginalized. Services are excellent for those who can afford them, and it is argued that a private, profit-focused system is better than no public transportation system."*

Figure 8-4. "Scenario 4: Prioritizing Social Cohesion," from Institute for the Future's The Future of Transportation: Five Scenarios: *"This transit system that focuses on social cohesion benefits almost all San Francisco Bay Area residents, but it also diverts resources away from routes optimized for efficiency or revenue. As well, people who wish to keep their neighborhoods insular are largely opposed. While a growing movement of concerned people has come forward with a list of reforms to rectify this and other issues, the small number of people who oppose the system overall use these issues as a wedge to cast doubt on the whole system."*

Figure 8-5. "Scenario 5: Embedding Regeneration," from Institute for the Future's The Future of Transportation: Five Scenarios: *"In this future, natural and rewilded environments win the most, with benefits for society as well. Metrics to measure and assess the regenerative economy are yet to be standardized, while the long-term impacts of rewilding cities remain uncertain. However, riders and underprivileged communities gain quality of life as a consequence, while providers of greener transportation benefit from a much larger market. Climate activism is at an all-time high, but eco-authoritarianism is also rising. Knowing when and what to choose among top-down and bottom-up strategies and management is key in this future."*

In a conversation I had with Jake Dunagan, director of the Governance Futures Lab at IFTF, he explained to me how he successfully used a very detailed image of a scenario to help stakeholders see multiple facets of a future world:

We worked with a major car company to develop scenarios for 2039. The written scenarios were good, but the real impact came from the visual images we created to represent them. The images were detailed scenes from four cities (Rome, New York, Rio, Paris) set in four alternative futures. At a high level, they captured the overall mood (misery, or high-tech wonder, etc.). But at a closer look, they were packed with details, so you could "zoom" in and see small but significant changes from the present. For example, you could see little details like a clock with a temperature on it, and there were fires on the hill; everything had a story tied to it. For a two-dimensional image, it allowed you to do a bit of telescoping to focus in on the details. What our clients did was take those images, and they projected them really large on the wall whenever they met with their teams and their executives, and I think it had a really powerful effect. This was the real payoff, because within less than three months after we delivered that project, the client made a big announcement saying that they were going to move their carbon neutral initiative goals from 2050 to 2039.

UN DPPA Science Fiction Posters

In 2020, the United Nations Department of Political and Peacebuilding Affairs (UN DPPA) commissioned Ronan Le Fur (aka Dofresh), an artist at Netflix, to create several futuristic posters for the 75-year anniversary of the UN (Figures 8-6 and 8-7). Le Fur collaborated with Martin Waehlisch and Benjamin Løzninger at the UN to reimagine the future of peacemaking by developing images of the future as an addendum to a global competition and celebration titled "Futuring Peace." Each poster highlighted a different scene in the peacemaking process—for example, negotiations in the general council room, analysts conducting field operations in the middle of a desert, a celebration and televised peace accord agreement, and a discussion aboard a military transport vehicle. Each scene featured hints of future technology, including robots, VR, holographic privacy bubbles, and drones. But what was key for Waehlisch and

Løzninger was to include many of the mundane objects that the UN still uses today, like coffee mugs, water pitchers, ordinary chairs, and paper, to highlight that the future, no matter how laden with technology, always includes everyday objects. Today, the posters are used in workshops around the world as provocations for leaders, ambassadors, and field operations personnel to think about future technologies, the gaps and aspirations, and the opportunities and cautions around adopting future tech into their lives. UN employees are asked to analyze the scene and the potential scenario they are in without being given any context and then are encouraged to have a discussion about the future of peace operations, where they could employ technologies, and how this future should or should not be augmented.

Figure 8-6. UN DPPA Innovation Cell sci-fi poster: analysts conducting a field operation

Figure 8-7. UN DPPA Innovation Cell sci-fi poster: a future peace celebration

Among the benefits of using an image or images in describing a future scenario are that they can:

- Provide a visual for a narrative
- Breathe life into a scenario when combined with animation and narrative
- Be powerful for discussion and debate
- Be useful for communication and marketing purposes when reporting out

There are also cautions associated with this approach:

- Depending on how you are using the illustrations, too much detail might seem overwhelming and overdesigned, while illustrations with too little detail could feel underwhelming.

- If overly detailed, the future could feel too fixed and not leave room for discussion or debate.

- If not illustrated or animated well, the images could seem under-developed and caricatured, suggesting that they should not be taken seriously.

While any type of visual can play a large part in helping an audience understand the future, *moving* images can provide even richer context. Futures is sometimes associated with science fiction (literature, film, and television), and the genre has served as the basis for many of the innovations we have today, but film and television are also long-form visual narrative depictions of scenarios. And while some may not take science fiction seriously as a viable source for inspiration in a corporate or business environment, it might be good to remind your audience that the iPhones, iPads, and AI they are using today began in the mind of visionary science fiction writers.

Science Fiction as Scenarios

Ernest Cline's sci-fi novel *Ready Player One* (Crown) was first published in 2011 and was adapted into a film directed by Steven Spielberg in 2018 (Figure 8-8). Set in 2045, the book portrays a world plagued by an energy crisis, pollution, global warming, overpopulation, and other issues that have led to widespread social problems such as famine, poverty, and economic stagnation. Many people escape their dystopian reality by immersing themselves in the OASIS (Ontologically Anthropocentric Sensory Immersive Simulation), a virtual reality entertainment universe (a metaverse). In this world, the OASIS is where everyone goes for entertainment, education, and to exercise and shop.

According to the story's synopsis on Wikipedia (*https://oreil.ly/_ky-B*), the OASIS

> *functions both as an MMORPG (massively multiplayer online role-playing game) and as a virtual world, with its currency being one of the most stable in the real world. It was created by James Donovan Halliday, founder of Gregarious Simulation Systems (formerly Gregarious Games), who made a posthumous video of his will stating to the public that he had left an Easter egg inside the OASIS, and the first person to find it would inherit his entire fortune, ownership of his corporation as well as complete control of the OASIS itself, which is worth trillions. The story follows the adventures of Wade Watts, starting about five years after the announcement, when he discovers one of the three keys which unlock three successive gates leading to the treasure.*

Figure 8-8. Tye Sheridan in the movie Ready Player One, *Steven Spielberg's adaptation of the best-selling novel by Ernest Cline (source: Jaap Buitendijk/Warner Bros.)*

Cline's description of the OASIS doesn't sound too different from China's social super app WeChat (*https://oreil.ly/YRYzz*), which combines multiple services in one place, including "text messaging, hold-to-talk voice messaging, broadcast (one-to-many) messaging, video conferencing, video games, mobile payment, sharing of photographs and videos, and location sharing." WeChat has dominated China for years and has even inspired US tech giants like Elon Musk[1] and Mark Zuckerberg's Meta (*https://oreil.ly/OISjJ*) to begin developing platforms that can provide every service you would need in your daily life with the intent of locking you into a single unified ecosystem. However, the implications of *Ready Player One*'s story are much darker, as the users of the OASIS's services become addicted to living out their lives (and fantasies) in VR rather than in the real world. People play themselves into poverty, family relationships dissolve, and authoritarian tech companies become the political forces that drive the economies of the world. IOI (Innovative Online Industries), a global telecom conglomerate and internet service provider and the story's villain, "encourages OASIS fans to run up debts that it collects by forcing them into indentured servitude."[2] This dystopian novel is riddled with analogies to 21st-century social media and cell phone addictions, virtual economies, and cryptocurrencies, as well as visions of how Web 3.0, VR, and the metaverse could play out if not regulated or mediated safely.[3] It is a cautionary tale, a testament to contemporary belief systems and behaviors, and a speculative scenario based on the trends of Cline's era, but *Ready Player One* is just one version of the future that could arrive, though it's a valid scenario that is shocking and entertaining to experience.

---

1 On October 27, 2022, Elon Musk acquired Twitter with the intent to rebrand and redesign the app (now called "X") into a super app with a similar variety of services as WeChat.

2 A. O. Scott, "Review: Spielberg's 'Ready Player One' Plays the Nostalgia Game" (*https://oreil.ly/iJkQk*), *New York Times*, March 28, 2018.

3 Web3 (also known as Web 3.0) is an idea for a new iteration of the World Wide Web that incorporates concepts such as virtual reality, metaverse, artificial intelligence, decentralization, blockchain technologies, and token-based economics. Whereas Web1 was characterized by static web pages and Web2 is characterized as "the web as platform," Web3 is slated to transform how people access the internet and how businesses operate in an attempt to undo the monopolization of large tech companies that have controlled so much of the world's data and content in the early 2000s. In essence, Web3 is poised to be the next evolution of the internet, one that detaches users from static pages and provides a new way to browse and navigate the world through advanced digital services and platforms.

The great thing about narratives and time-based media such as film is that they give the author or creator more time to build out the world for the reader or viewer. Rather than everything being presented on one page, the world is wrapped within a slowly unfolding story about people with emotions, consequences, and opportunities that impact their lives in sometimes very opportunistic and detrimental ways. The use of drama in scenario development can be very powerful when you want to create a belief system that pulls on the hearts and minds of your audience, creating an imperative to co-imagine that world and activating their sensitivities about a world they may or may not want. Unfortunately, much of science fiction is designed to create drama through hardship and dystopian themes. It's a device that keeps people engaged and wanting to see the story's resolution. This isn't necessarily a wrong tactic, and in fact it has inspired many over the years to visualize futures that could go awry. In the wake of so many films like *War Games*, *Blade Runner*, *The Terminator*, and *I, Robot* that depict how AI or robots could try to destroy humanity, and as AI and robots begin to evolve and proliferate the world today, tech leaders and governments are now banding together to design policies to regulate the advancement of AI so it doesn't go down the dystopic paths depicted in those movies.[4]

IMMERSIVE AND INTERACTIVE EXPERIENCES

Kaspersky's Earth 2050 (*https://oreil.ly/3iEUG*) is a 3D animated website that allows you to see different futures in 2030, 2040, and 2050 across different cities around the world (Figure 8-9). The interactive site allows you to click into each city, select a future technology or service, and learn more about it. It also includes a public chat and forum so that people can leave comments, approve or disapprove of the prediction, or engage in discourse and debate about the ideas presented. It's a fun and interactive way to see how futures could evolve across different decades and in different regions of the world (Figure 8-10).

---

4 The European Union Artificial Intelligence Act is one example of these policies; see Martha Buyer, "The EU Takes the Lead in AI Regulation—but the New Rules Will Have Global Implications" (*https://oreil.ly/0Bpz0*), *No Jitter*, December 20, 2023.

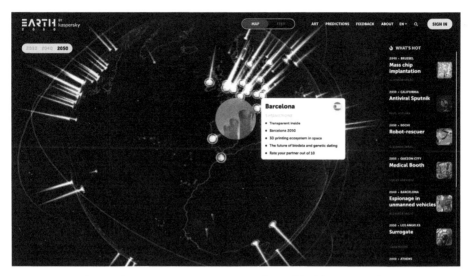

Figure 8-9. Earth 2050 home page: visitors can rotate the model of the earth and select a specific city they want to explore

Figure 8-10. Earth 2050—Shanghai in 2040: selecting a certain city reveals an immersive 3D environment that the users can drag and rotate to see a 360-degree view of the city. Certain technologies or "predictions" are also highlighted, and users can click them to find out more information.

In 2006, Stuart Candy and Jake Dunagan, colleagues at the Hawaii Research Center for Futures Studies (HRCFS), conducted a large experiential workshop for the city of Honolulu called "Hawaii 2050." In a 2016 blog post (*https://oreil.ly/dneVh*), Candy described Hawaii 2050 as "an initiative of the state legislature to engage the public in addressing the islands' sustainability, over a commendably farsighted time horizon, nearly two generations out. At the time the project began, in late 2005, more than three decades had passed since the last comparable process. A big-picture re-examination of Hawaiian prospects was overdue." At the workshop event, Candy and Dunagan put 550 people into 4 different rooms, each representing a different scenario of the future (Figure 8-11).

Figure 8-11. Some of the participants in the Hawaii 2050 Immersive Futures workshop organized by Stuart Candy and Jake Dunagan for the Hawaii Research Center for Futures Studies (source: Stuart Candy/Hawaii Research Center for Futures Studies; photos by Cyrus Camp)

Candy's look back at the workshop continued:

Our intention was to give people a chance not just to contemplate these potential realities as intellectual hypotheticals, but to visit physically and invest emotionally in them. Thus this set of brief yet provocative immersions, instantiating highly contrasting assumptions and theories of change, as a fast-track to higher quality, more richly imaginative mental models and civic conversations....

The "Hawaii 2050" scenario rooms were therefore immersive and experiential not only literally, with participants surrounded by performers and designed media (set dressing, props, soundscapes, etc.), but also narratively, with each room devised as a coherent scene that would place attendees in medias res, *and invite in them a sense of being transported in time. We aimed to ensure that people would not only witness these futures firsthand, but interact with and within them[;] hence, four situations from alternative futures: a pre-election speech night; a naturalisation ceremony for climate change refugees; a government-mandated sustainability class; and an information session at a posthuman wellbeing facility.*

Stuart and Jake's pioneering work in experiential futures eventually led to a framework called the *Experiential Futures Ladder*, which frames the different layers necessary to transport audiences into the future so that they can experience future worlds. (I'll discuss the Experiential Futures Ladder in more detail in Chapter 10.)

So how do you decide which scenario to focus on? This decision can be overwhelming to people who have never worked with scenarios before. Try to manage expectations early about what scenarios are, how they'll be used, and why we need to sit for a moment in each world before we make any grand decisions. After considering the options, you may want to converge on one scenario to work with. Maybe that scenario is a collection of characteristics from other scenarios, or perhaps it's a very specific one. Because the future really is a combination of many scenarios—it is dystopic to some and transformational for others—the decision about which scenario to work with depends on what you are focusing on and what you do with that scenario. Be careful not to lean too heavily on any one type of scenario (unless that's the purpose of the project), or else you might skew people's perceptions of what the future looks like by disregarding additional factors that could influence change.

There are several benefits to the use of immersive and interactive experiences to portray future scenarios:

- It allows you to investigate various polarized themes of future worlds.
- It can show a holistic view of future environments (with STEEP).
- It can portray a day-in-the-life of certain actors or places in the future.
- It allows you to explore alternative situations and prototype new ideas.

- It facilitates deeper thinking about dependencies and various dimensions of future worlds.

As you would expect, there are some cautions as well:

- Polarized themes may seem extreme.
- Too many or too few scenarios could ignore important factors, dependencies, and conditions.
- Depending on the format, you may need to include very specific details, devices, or information to prepare and immerse your audience so that they can suspend reality and truly be in your future world.

Surviving the OPEC Crisis with Scenarios at Shell

One of the most famous examples of scenario planning comes from the Royal Dutch Shell Oil Company, which has been using scenarios since the 1960s. The process was first introduced in 1966 in the organization's exploration and production (E&P) function, "where some considered that it was desirable to try and assess the political risk of upstream investments by seeking to understand the forces at work."[5] However, it was Pierre Wack, head of the Shell Group's planning scenario team for most of the 1970s, who was known for bringing scenario planning to executive management and really challenging traditional management thinking. Often called the "Father of Shell Scenarios," Wack urged executives to abandon their assumptions, including the global belief that oil supply was infinite. Among the accomplishments attributed to Wack and his efforts to integrate scenario planning were the prediction of the oil crisis in 1973–1974, when OPEC (Organization of Petroleum Exporting Countries) imposed an embargo on the United States in retaliation for

---

5 Michael Jefferson, "Shell Scenarios: What Really Happened in the 1970s and What May Be Learned for Current World Prospects," *Technological Forecasting and Social Change* 79, no. 1 (January 2012): 186–197.

the US decision to resupply the Israeli military during the Arab-Israeli War. Scenarios allowed Shell to anticipate the evolving challenges in the Middle East, including dynamically shifting political issues. Wack's envisioning of what he called "the rapids" depicted the oncoming oil embargo by OPEC against the US and other countries, as Middle Eastern countries began to nationalize their oil assets to leverage more control over the global market (or political adversaries).

> There was serious failure up to early 1974 to understand fully and take due note of forces "already in the pipeline"; failure to draw on past experience—from economic and social history, from past financial crises, from the operation of business organizations and industrial cartels, from past military and religious conflict; from the history of the Middle East (especially the Gulf); and failure to assess sufficiently the risks of taking one's eye off the short and medium term for whatever reason.
>
> —Michael Jefferson, "Shell Scenarios: What Really Happened in the 1970s and What May Be Learned for Current World Prospects"

> The central purpose of Shell's scenario work was, however, in the memorable words of its Head of Business Environment—Pierre Wack—"to shift the personal microcosms" of Shell's employees. And this it did for many in those early years. But it is one thing to move people away from "Business as Usual" attitudes to under-standing and working with the numerous forces which may cause "perturbations." It is not only a question of looking at future pos-sibilities, but of understanding the past (the failure to do so has had remarkable repercussions in the financial and banking crisis of the past three years).
>
> —Michael Jefferson, Centre for International Business and Sustaina-bility, London Metropolitan Business School

The 1973 scenarios first established Shell's reputation for using this hitherto academic approach to inform strategic business planning. They helped Shell weather the volatility of the 1970s, bringing financial gains running into the billions of dollars thanks to the sale of refineries and installations or decisions not to replace them. Shell has spent four decades since then producing and using many types of scenarios to anticipate global economic, social and political changes and their likely impact on business. Summaries of some of these have been regularly shared outside Shell, contributing to important public debates.

—Jeremy Bentham, head of scenarios, strategy,
and business development, Royal Dutch Shell

By the 1980s, the repercussions of global shifts in oil production and the economy were being seen by Shell's supply and marketing function in assessing crude oil production and price prospects. And though the late 1970s into the 1980s saw lots of turbulence, it was this work, along with the adaptation of medium-to-long-term scenario planning, that allowed Shell to compensate and survive the rapid changes well into the 1990s. While the story was not as simple as that sounds, and the kinds of scenarios that Wack presented were sometimes met with contention by leadership, his scenario planning allowed Shell to have an organized approach to dealing with the drastic oil price increases due to the embargo and other market factors and the embargo's impact on imports and exports globally, and to develop a strategy that would become a mainstay and a case study for decades to come.

Scenarios in general are one of the key tenets of Strategic Foresight. It will be important for you to think about how to use these in your process and journey. The format within which you present the scenarios is just as important as the scenarios themselves. In fact, it is said that Pierre Wack's work at Shell was so impactful due in part to his ability to package story lines, and because he was also an incredibly inspirational presenter. Think about how much detail is necessary and what might be needed (from your diagnostic) to make scenarios work. Consider also how much you will have to immerse your audience in these worlds to make certain points about the gravity of each situation and why they should be taken into consideration. You may even want to include short- and medium-term scenarios (like Shell did) to connect the near and far futures by providing building blocks toward those futures. Creativity, facilitation, storytelling, and format will be key in making this a useful artifact in your process.

Jim Dator's Scenario Archetypes

When deciding whether to introduce scenarios as a concept or a methodology, depending on the audience, you might consider starting with an explanation of the four traditional archetypes of scenarios developed by Jim Dator in 1979. Dator, a political scientist and futurist who served as the director of the PhD program in Alternative Futures at the University of Hawaii at Manoa, presented a pioneering framework for societal change in the book *Perspectives on Cross-Cultural Psychology* (Academic Press). Dator's framework suggested that the common depictions, narratives, and scenarios of the future could be categorized into four scenario archetypes (Figure 8-12). The labels assigned to these categories were *continuation, collapse, transformation,* and *disciplined society.* While the original archetypes operated at a broader societal perspective, they served as the foundation for how many use scenarios in foresight today:

Continuation (business as usual)

The great joke within futures is that the continuation scenario is not realistic, but we still need to consider it, because if there is little to no change in this future, that means the future not only inherits today's conditions but also inherits its problems. In a continuation scenario, nothing is necessarily solved or transformed, which means little innovation or evolution of thought or process, and thus we continue on a similar path to the one we are on today. While this seemingly "conservative" approach might feel safe, it doesn't necessarily progress our thinking, foster innovation, or recognize a rapidly changing environment or evolving competition.

Collapse

The collapse scenario is fodder for many science fiction narratives. This is a dark future in which systems and society collapse. That may feel a bit dramatic, but you can also consider the collapse scenario as a nonpreferable, least-ideal, or negative outcome. Without collapse scenarios, we aren't prepared for the worst and risk more by not interpreting issues that could eventually lead to failures or adverse situations.

Transformation

The transformation scenario (not to be confused with a utopia, where everything is perfect) is a world in which growth, evolution, and innovation thrive all around us. It is the epitome of the preferred "happy path," where we solve problems and continue to grow and develop as a society and species. Transformation scenarios present the "best cases" but don't necessarily have to be perfect, shiny, and happy. They map out a path in a positive direction so that we can assess its implications.

Disciplined society

The disciplined society or discipline scenario is one in which society continues to adapt to changing conditions. It recognizes disruptive events and attempts to save itself and return to a new equilibrium. How society responded to the COVID-19 pandemic is a good example of a disciplined scenario. Due to the pandemic (a collapse event), society had to compensate to survive. A new equilibrium was established by changing how we live, work, eat, and play. Society's values and operations shifted to maintain a discipline and order that we all craved. We worked from home, we ate more local food, and we found other sources for our supply chains in an attempt to regain a different type of normality.

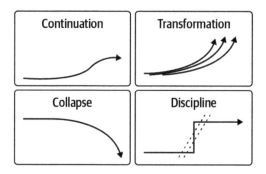

Figure 8-12. Jim Dator's Four Futures (scenario archetypes)

Dator's archetypes set the stage for different ways the future can play out. While they may seem extreme, dramatic, or limiting, they provide a baseline for how to frame or assign themes to the worlds we are envisioning. We don't necessarily need to use such extreme labels as "society collapses" or "society

transforms"; the scenarios can be framed more simply as "things go well" or "things don't go well." However, how you frame or depict these scenarios for your team, client, or audience can be key to helping them understand the gravity of the world you are painting for them. Saying a collapse scenario is "the business goes under," "economic downturn," or "thriving competition" could work to your advantage by stoking feelings and emotions about the "things don't go well" scenario. And sometimes that type of marketing language is necessary to transform how people perceive impending threats or opportunities. Dator's archetypes have served as a foundation for scenario planning, but they aren't the only way to frame futures. Nikolas Badminton, in his book *Facing Our Futures* (Bloomsbury Business), refers to positive and dystopian futures as a way to position opposing scenarios in his Positive-Dystopian Framework:[6]

Positive Futures—*a world where we have a global view, and infrastructure to support, improving health and wellness, and reducing wealth disparity. A world where we design humanity-centric, balanced and egalitarian solutions to our greatest challenges. A world where we rely on a reduced reliance on the few companies that want to change the world and towards an empowerment of every person on the planet to have ownership of their own identity (and their data), and a right to exist with their hopes, dreams and opinions in any place in our world, and in our futures. A world where we see infinite futures and plan for them accordingly through the application of positive principles of change. And, ultimately, a world where wars are no more, and any conflict can be resolved through rational discussion and collaboration.*

Dystopian Futures—*a world where we perpetuate the reliance on the industrial complex, supported by military action and conflict, and where billionaires and shareholders are rewarded ahead of any of the users of technology solutions/platforms. A world that is led by short-term thinking and greed, and where personal protectionism is apparent and unashamed. A world where leadership ego is left unchecked and short-term effects are rewarded. A world of corporate and governmental surveillance and control to ensure we all have futures that fall in line with the visions of the few people that want to shape the singular future.*

---

6 Nikolas Badminton, *Facing Our Futures: How Foresight, Futures Design and Strategy Creates Prosperity and Growth* (London, UK: Bloomsbury Business, 2023), 122.

When we create a user interface, we have to ask ourselves several times over: what if the user does this or that? We consider those conditions and create ways the system can respond to those scenarios. We can present error state language in a pop-up window or redirect users to a different page. In quality assurance (QA) testing, we test every kind of scenario we can think of to make sure the system doesn't break, harm, or inconvenience our users. But in the assessment of which scenarios might appear, there are definitely some that are imminent (bound to happen) and others that are *edge cases* (could happen but maybe not often). For some edge cases, we might design a response for the rare occasion in which the scenario does happen, while for scenarios that are more likely to occur, we may build more complex responses or even conditions that will prevent the scenario from happening (a safeguard of some sort).

It is much the same with scenario planning. Some scenarios are more probable or plausible than others. With enough investigation and analysis, we could add a likelihood index that certain scenarios will in fact appear and thus we might want to prepare for them. If you are sailing in the ocean and know (through radar forecasts) that a storm is approaching, you might want to buckle down and prepare for it by enacting safety procedures. But if the chances are low that the storm will appear, you may just want to be aware of it, and if it does cross your path, you'll know what to do. We can use the same mentality when designing scenarios and deciding how we might want to respond to them later. We have to consider different types of events (both good and bad) and think about them deeply so that we can decide which are the most important ones to consider when planning our strategies.

The Axes of Uncertainty

The future is laden with uncertainties. Remember: VUCA (volatility, uncertainty, complexity, and ambiguity) is around every corner and can reconfigure or reassemble at the drop of a hat, creating more and more uncertainty. Just when you think you are prepared for one thing, another event surprises you. Scenarios alone won't tell you everything that could happen. And it can be quite maddening when you realize just how many alternatives could exist. Luckily, foresight has a number of tools for assessing and diagramming uncertainties. One such tool is the *Axes of Uncertainty* (Figure 8-13).

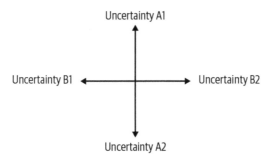

Figure 8-13. The Axes of Uncertainty

To use this tool, create a 2×2 matrix and label each axis with opposing views or "headlines" of a particular topic or uncertainty. Each quadrant becomes the intersection of the topics, which can be a way to understand how opposing topics might collide. Once you begin to experiment with this framework, you'll start to see some similarities in how it can polarize different outcomes (much like scenarios do). In fact, you can essentially use the framework to begin plotting details and themes for your scenarios. Due to the opposing nature of the axes, they naturally create four quadrants, which could easily be mapped to the four scenario archetypes.

We tend to under-predict or over-predict change. The reason: imagining plausible outcomes forces us to confront our expectations and cherished beliefs. If you were to jot down descriptions of the future, you'd quickly find that they mirrored your own cognitive biases. You'd focus mainly on the subjects you already know well—like your work, your company and industry—so you'd miss all of the related risk and opportunity ahead.

—Amy Webb, Futurist, The Future Today Institute

While the axes can be based on very specific factors, this tool does present a way of looking at different angles or outcomes of a subject you're investigating. You could use several of these diagrams to place more definition around a scenario you are trying to develop based on varying conditions (Figure 8-14).

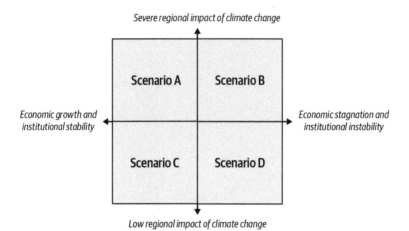

Figure 8-14. Using the Axes of Uncertainty to analyze the regional impact of climate change versus economic growth and institutional stability

The benefits of using the Axes of Uncertainty are severalfold:

- You can speculate on uncertainties by forcing a polarity on themes.
- It gives you a method to initiate scenario development.
- It allows you to plot very different themes, ideas, or headlines against each other.
- It can enrich scenario work by helping you to see how other factors could influence or impact a particular theme.

The cautions associated with this tool are also severalfold:

- You are limited to only two opposing topics at a time.
- The axes can seem like they are limiting or lack depth.
- If using them to start or build scenarios, you must recognize that they are based on limitations or other dependencies, so you need to fill in gaps or do multiple axes exercises to flesh out a more robust scenario.
- While the axes can lead to scenario development, scenarios are made of many factors, not just two opposing criteria; be sure to consider that there are multiple factors involved in future worlds. However, several axes exercises can lead to the creation of more robust scenarios.

Future Personas

In traditional Human-Centered Design, we aggregate users or customers into generalized characters or archetypes called *personas*. They act as the guidelines for what we are designing and for whom. Futures Thinking operates a little differently. And for those of you who might be programmed to look at everything through the user's eyes first, you may have to step out of your comfort zone as you read this section. While you can begin your research with a particular persona or target market, demographic, or segment of a population, you can also develop personas in more detail later on as you begin to describe future worlds through scenarios. And though foresight doesn't necessarily omit people from the research, people (and systems) are inherently built into the trends that you research since they are some of the conditions that drive them. Identifying early on who you might be interested in investigating in the future is useful for focusing research efforts at the beginning. This allows you to gather some data on the people within a region so you can understand growing social/cultural trends and how they might be driving other trends into the future. But there is still much to learn about the future scenarios they might be living in. Sometimes you may want to develop your future personas *after* you write scenarios (Figure 8-15). This allows you to situate the personas into more specific situations to understand how they might behave differently in each scenario should it arrive. For instance, a collapse scenario such as a pandemic will drive very different behaviors than would a transformational scenario in which no pandemic exists. We know this is true because we saw the stark differences in what we value, how we spend our time and money, and what products and services we used or needed during pandemic times. Today, as the pandemic is behind us, our lives are very different. Our personas and other actors in the world are driven by a completely different set of values and conditions. Thus we can see how scenarios can truly change how personas are described and used.

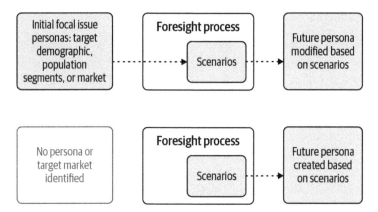

Figure 8-15. Starting with a target persona and modifying it into a future persona after scenarios versus creating a future persona from scratch after scenarios

In his paper "The Future Persona: A Futures Method to Let Your Scenarios Come to Life," Alex Fergnani, a foresight practitioner, consultant, and educator, states that[7]

> *scenarios enriched by future personas viscerally connect with an audience [Figure 8-16]. Be it a group of corporate executives who needs to be convinced of the importance of taking action in view of likely future disruptions, or a large community of individuals who needs to be sensitized on a more sustainable course of action in light of an emerging issue. By reading the actions, thoughts, missteps and tribulations of a future persona, we feel as if we are living the scenario ourselves. We feel as if the persona is a real human being, we identify with the persona, and we judge the actions of the persona.*

A future persona can have all the same attributes as a traditional persona, including, behaviors, emotions, needs, and barriers, but it is influenced by future conditions. You could have a future persona for every scenario you create, each one behaving a little differently based on its scenario (Figure 8-17).

---

7 Alex Fergnani, "The Future Persona: A Futures Method to Let Your Scenarios Come to Life" (*https://oreil.ly/UhTFR*), *Foresight* 21, no. 2 (June 2019).

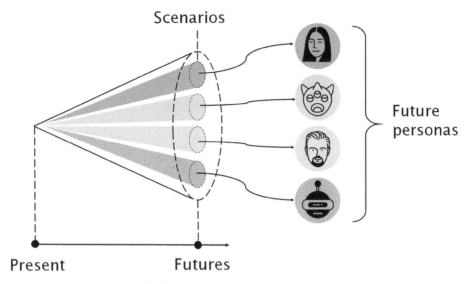

Figure 8-16. Future personas by Alex Fergnani

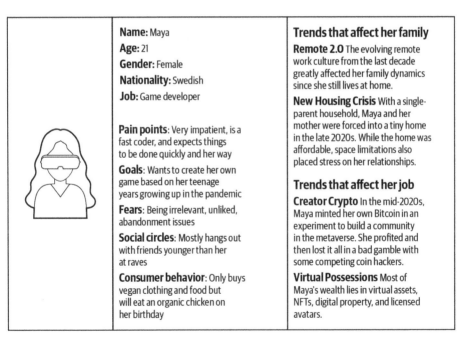

Name: Maya
Age: 21
Gender: Female
Nationality: Swedish
Job: Game developer

Pain points: Very impatient, is a fast coder, and expects things to be done quickly and her way

Goals: Wants to create her own game based on her teenage years growing up in the pandemic

Fears: Being irrelevant, unliked, abandonment issues

Social circles: Mostly hangs out with friends younger than her at raves

Consumer behavior: Only buys vegan clothing and food but will eat an organic chicken on her birthday

Trends that affect her family

Remote 2.0 The evolving remote work culture from the last decade greatly affected her family dynamics since she still lives at home.

New Housing Crisis With a single-parent household, Maya and her mother were forced into a tiny home in the late 2020s. While the home was affordable, space limitations also placed stress on her relationships.

Trends that affect her job

Creator Crypto In the mid-2020s, Maya minted her own Bitcoin in an experiment to build a community in the metaverse. She profited and then lost it all in a bad gamble with some competing coin hackers.

Virtual Possessions Most of Maya's wealth lies in virtual assets, NFTs, digital property, and licensed avatars.

Figure 8-17. Example of a future persona

Personas don't necessarily need to be human either. A new field of design called *Life-Centered Design* focuses on nonhuman actors, including animals, plants, environments, nature, and even objects and buildings.[8] As an antithesis to Human-Centered Design, Life-Centered Design considers all the other actors in an ecosystem *before* humans' needs. Future personas could also include nonhuman actors when thinking about scenarios' implications in future worlds.

Sarah Owen, Global Futures Advisor at Soon Future Studies, explains how they use personas:

> What we'll do is we'll look through the lens of who that human will be in the future. And so it'll be a day in the life of that consumer persona, and through social context as well. So we situate that quadrant through the lens of what society and culture looks like. And then our version of a persona is called "A Day in the Life," and we'll take the client on a journey so they can see the different touch points that someone would engage with, from what they listen to, how they commute to work, or how they communicate with friends.

The use of future personas brings several benefits:

- They bring scenarios to life by highlighting actors within them.
- They create a richer perspective of impact on future human (or nonhuman) needs.
- They can be used to depict or understand nonhuman or human actors.
- They can detail a day in the life of future actors.

---

8 As defined in a 2022 *Medium* article by Damien Lutz (*https://oreil.ly/tNW47*), Life-Centered Design is "an emerging design approach that expands human-centred design to also include consideration of sustainable, environmental, and social implications. It connects micro-level design (UX, product engineering, etc.) to global goals by increasing the stakeholders from just 'user and business' to 'user, non-user, local and global communities, ecosystems, and planetary boundaries.'"

There are also cautions to consider, of course:

- Using future personas could feel too speculative, since we can't actually talk to people in the future.
- Human-centered designers could struggle with doing this exercise later rather than sooner in the process.
- It could feel like an afterthought in the process rather than a priority.

Implication Mapping

Since scenarios are collectively a result or projection of current trends and patterns, they are also the consequences and implications of those trends. And any trend can go in a number of directions, depending on many factors. Another way of extrapolating the potential impact, influence, or evolution of a trend, event, or signal is to think about it as a cause-and-effect model. If a trend is what it is today, what are the different ways it can impact the world or people over time? And how does that continue to play out over several years if it keeps changing and affecting other trends, technologies, cultures, economies, or policies? What we might want to do to expand our analysis is to map out the variety of implications so that we can get a wider view of how a trend can play out into the future. We can do this with a method called the *Futures Wheel* (aka the Implications Wheel or the Wheel of Implications). Depending on the audience, I might call it something different, but I always describe it as a way to investigate impact scenarios or consequences. For designers, I may use the term "Futures Wheel," whereas I may call it an "Implications diagram" for corporate executives. Modifying the title of the exercise can sometimes be useful in shaping how people perceive what it is, what it's used for, and how they participate in creating it. Personally, I believe the Futures Wheel is one of the most useful and flexible tools in the futures toolkit. It seems to be the one method that is fairly easy to explain and fun to use, and it can provide a multitude of valuable insights, perspectives, and ideas. Whether you're working on a 10-year business strategy, creating new products, or planning a wedding, the Futures Wheel is just another way to conduct an opportunity and risk assessment of potential consequences.

Invented by Jerome Glenn in 1972, the wheel allows you to map out multiple branches and threads of consequences that could emerge as the result of a central topic. Similar to a concept map or mind map, it uses circles (as content bubbles) and lines (as spokes) to connect one thought (or impact) to another (Figure 8-18).

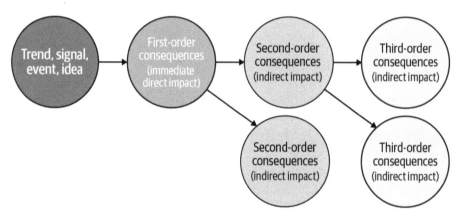

Figure 8-18. How to create a Futures Wheel

To create a Futures Wheel, draw a circle, and inside it write your central topic—this can be a trend, event, signal, change, or idea (it can be anything you want to investigate, really). Then draw another circle outside of the first circle. This is where you write the direct impact of this topic; it is usually the most obvious consequence you can think of. Let's say your topic is the advent of autonomous (driverless) cars. One direct implication of owning an autonomous car is that you don't have to drive anymore. Another direct implication is that roads might be safer. Draw all these impacts as branches extending from the central topic. These direct implications can also be labeled as *first-order implications* or *first-order consequences*. You can then draw another circle extending out from each first-order implication. These will be the *indirect impacts* or *second-order implications* or *consequences*. In our example, the first-order implication is "You don't have to drive anymore," so a second-order implication or consequence might be

"Since we don't have to drive anymore, we might forget how to drive in the event of an emergency." This impact is essentially the result of the first implication (if *this* happens, then this other thing might happen). For each implication you can draw another implication and extend further out from the central topic in second order, third order, fourth order, and so on. A third-order implication of "We might forget how to drive" could be "If the car has an error or fails, we could be injured since we forgot how to drive" (Figure 8-19). The third-order implication may feel unlikely if you are investing a lot of time and money into building an autonomous car that is supposed to protect lives and make the roads safer. But what if? What if an error does occur? Because errors do occur, and if you don't have a contingency plan, you might pay the consequences.

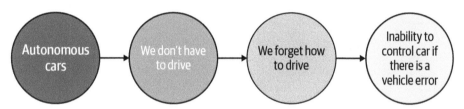

Figure 8-19. Futures Wheel implications of autonomous cars

The Futures Wheel allows you to look at many potential outcomes, just as scenarios do (you could even label certain implications as general "scenarios"). Now that you've uncovered a potential threat (or opportunity, depending on how you look at it), you'll need to decide how important it is to address it. Do you want to design a solution for this implication? Or is it enough simply to know it *could* happen and have a conversation about it? You may want to consider this seemingly "edge case" condition and create a fail-safe for it, because if it does occur, you could lose (or save) a life. How important is it to you now?

Eventually, a completed Futures Wheel diagram will look like a giant spider web, with the different orders of implications radiating from the center (Figure 8-20).

Figure 8-20. Futures Wheel

"Since we don't have to drive anymore, we might forget how to drive in the event of an emergency." This impact is essentially the result of the first implication (if *this* happens, then this other thing might happen). For each implication you can draw another implication and extend further out from the central topic in second order, third order, fourth order, and so on. A third-order implication of "We might forget how to drive" could be "If the car has an error or fails, we could be injured since we forgot how to drive" (Figure 8-19). The third-order implication may feel unlikely if you are investing a lot of time and money into building an autonomous car that is supposed to protect lives and make the roads safer. But what if? What if an error does occur? Because errors do occur, and if you don't have a contingency plan, you might pay the consequences.

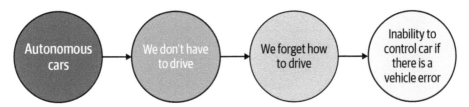

Figure 8-19. Futures Wheel implications of autonomous cars

The Futures Wheel allows you to look at many potential outcomes, just as scenarios do (you could even label certain implications as general "scenarios"). Now that you've uncovered a potential threat (or opportunity, depending on how you look at it), you'll need to decide how important it is to address it. Do you want to design a solution for this implication? Or is it enough simply to know it *could* happen and have a conversation about it? You may want to consider this seemingly "edge case" condition and create a fail-safe for it, because if it does occur, you could lose (or save) a life. How important is it to you now?

Eventually, a completed Futures Wheel diagram will look like a giant spider web, with the different orders of implications radiating from the center (Figure 8-20).

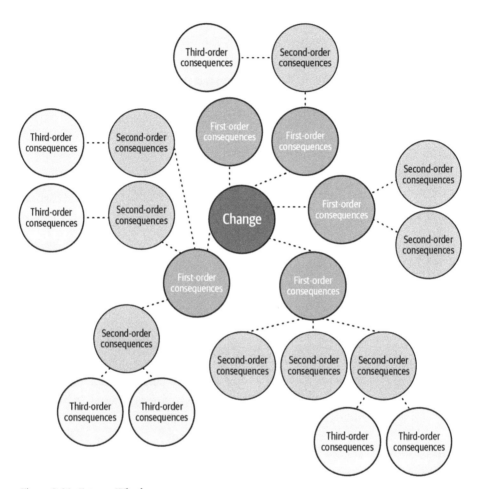

Figure 8-20. Futures Wheel

In February 2020, P A Martin Börjesson, a futurist in Sweden, created a Futures Wheel to map out the potential consequences of the coronavirus outbreak (Figure 8-21). While little was known about the virus at the time, and systems were beginning to be challenged or collapsing, Börjesson's Futures Wheel fairly accurately projected several implications that became a reality. Sure, there were many signals emerging that could easily be plotted onto first-order consequences. The second- and third-order implications, which were still speculative at the time, were logically derived from news and information we were seeing play out in real time around the world as the year progressed.

Figure 8-21. Futures Wheel on the coronavirus epidemic's longer-term consequences (source: Futuramb blog (https://oreil.ly/Gp-xv))

If we look more closely at one of the branches of the Futures Wheel (Figure 8-22), we can see some implications that seem obvious today, such as "Challenges to medical organizations, resources, equipment, supplies and people." That implication is obvious, as most first-order implications usually are (the most obvious impacts), but as you go farther out, you see implications that at the time were just speculation, such as "Fall of weak political leaders" and "International relations and trust is challenged"—all of which did actually happen during the pandemic.

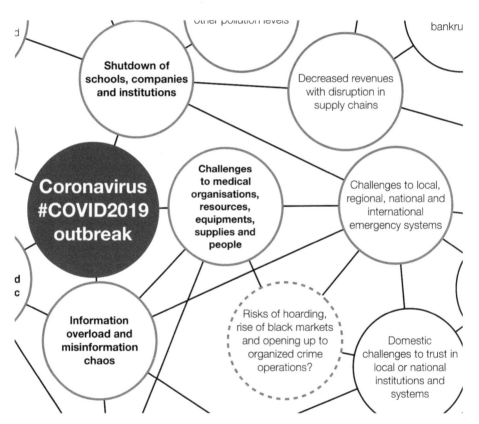

Figure 8-22. Detail from Futuramb's COVID-19 Futures Wheel (source: Futuramb blog (https://oreil.ly/Gp-xv))

The reality is that many people and organizations may only think about first- or second-order implications (more so the first-order ones), and that isn't necessarily wrong. But just by taking the time to go a few steps further and consider multiple types of impact, you can be that much wiser and more informed. You can also use STEEP (social, technological, environmental, economic, political) categories to organize or inspire your thinking. If you get stumped, sometimes shifting your perspective into one of the STEEP categories can spark new thoughts. For example, how does the use of generative AI for architectural planning impact society from only a *sociocultural* lens? Or how might it create a new *economy* for architects? Or what are just the *political* or *regulatory* implications of this technology or trend? Categories are just another way of applying a different impact lens on the topic and can generate new thoughts and ideas that you may have overlooked before. Each order of impact can also be categorized by a time horizon. For example, you could ask what implications could result from this event in the next six months, in one year, or in five years. This division, much like using the STEEP suggestion, is just another way of placing guardrails on your thinking, as well as a way to organize your thoughts within boundaries across time horizons.

Much like with any generative exercise, you'll eventually end up with a lot of perspectives. It will be important to prepare a prioritization exercise or a way to select which implications are most important for you to discuss or address. You can set up a quantitative method by assigning values to implications (impact to business, probability of occurrence), use the Probability Versus Impact Matrix or some other formula, or vote on which ones seem compelling, threatening, impactful, or questionable. Whichever way you choose, you'll want to figure out how to organize and isolate the information to make it useful for working toward your project goals. When trying to discover new transformational opportunities, you may want to look at impacts on the outskirts (third- or fourth-order implications), because those are likely to be consequences that people may not even have considered yet, thereby creating a ripe area for exploration, experimentation, research, or innovation work.

A list of the benefits of implication mapping will include the following:

- STEEP categorization allows you to cover multiple perspectives of impact and drive thinking in different areas.
- You can uncover a variety of implications (positive, negative, neutral).
- It stimulates discussion about implications that are usually overlooked.

There are cautions to consider as well:

- There is only one central topic per wheel.
- Information overload can confuse or intimidate people, especially when they are trying to prioritize which implications to address.
- This approach can generate an overwhelming amount of ideas and perspectives; make sure you have a prioritization method planned to select and narrow your interests.

Sequencing Methods

If you haven't already noticed, many of these methods are just frameworks for extrapolating what we know today into the future. They just do it in different ways. As noted, Futures Thinking is not linear. A scenario can have similar information to an implication wheel. A future persona can be described as an implication of a scenario. A quadrant of the Axes of Uncertainty can describe a facet of a scenario. Scenarios can inform Futures Wheels or vice versa. What you need to remember is that these are just tools that can be used in any order, in parallel or alone; they are vehicles to take you into the future so that you can look around and assess the variations of that world. Some may argue this, but the reality is that, due to the nonlinearity of futures work, there are no strict rules as to which method you *should* do first or last.

Nikolas Badminton proposes a framework in his book *Facing Our Futures* that he calls the *Positive-Dystopian Framework*; it details a specific approach for moving through the stages of futures work in 10 distinct steps (Figure 8-23). His approach provides a nicely structured method for using the methods while incorporating both positive and negative futures throughout the entire process, ultimately leading to strategic planning.

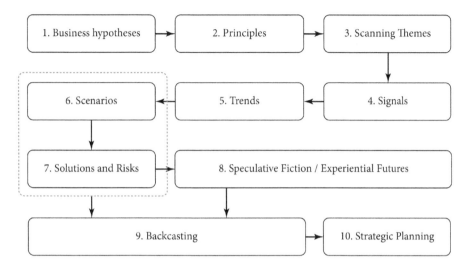

Figure 8-23. Nikolas Badminton's Positive-Dystopian Framework

The steps in Badminton's Positive-Dystopian Framework are as follows:

Step 1: Write the Business Hypotheses and Set the Horizons
We typically frame the work by outlining one to three hypotheses that we could be facing in our futures: business disruptions, societal tensions, war, climate change, mental health, and so on.

Step 2: Set Foundational Principles
Foundational principles provide the framing, rules, and guidelines that serve as a reference for thinking about the longer-term goals of our organization, how it plays in existing and new industries, and how it will be an active participant in shaping our future in the world.

Step 3: Scanning Themes
With the principles in place and a well-thought-out business hypothesis, we can start with a list of scanning themes/areas of interest to explore as we begin to scan for signals. Example themes could be food prices and logistics, global shifts in population and city growth, impacts of climate change, changing energy landscape, water scarcity, and other areas of interest.

Step 4: Signals
Signals are distinct pieces of information, technologies and digital infrastructure, geopolitical movements, cultural shifts, statistics, stories,

activities, and/or events that indicate an impending shift or change or an emerging issue that may become significant in our futures.

Step 5: Identify Trends

A trend is a general direction in which something is developing or changing; it is created by existing conditions and environments and is affected by one or more signals. After scanning for signals and ascertaining the positive and dystopian effects, identifying trends is one of the most important tasks when undertaking foresight.

Steps 6 and 7: Develop "What If?" Scenarios, and Identify Solutions and Risks

"What if?" scenarios help us create quick, short stories of possible actions or events in the future. These are short statement-based stories that present future states and provide the canvas for exploring specific areas. We then brainstorm what speculative solutions could exist, and we identify a time in which to state the scenarios.

Step 8: Speculative Fiction and Experiential Futures

Speculative fiction is the approach of writing fictional stories or creating visual elements that tell a story. These are typically an evolution of the scenarios we've developed with expressions of future artifacts. Experiential futures expand to include something we can touch, hear, walk through, and/or immerse ourselves in. Both forms may take people into supernatural and other imaginative realms.

Steps 9 and 10: Backcasting, and Strategic Considerations

Backcasting allows us to travel backward from the futures we have created to where we are today by identifying strategic considerations from the futures scenarios that we have built; considering what people need to be involved, what processes need to evolve or be installed, what governance (internal and external) is needed, and what solutions need to be put into place; and outlining the programs and projects that need to be established, as well as a view on the investments needed and who will help support that journey.

Badminton recommends that Steps 1 to 8 be your minimum, while Step 9 (Backcasting) and Step 10 (Strategic Planning) are additional steps that he recommends teams follow to help bring futures narratives and the relevancy of signals and trends, and their long-term effects, back to the discussions around strategic planning today.

This isn't to say that omitting a step or two will prevent you from creating a successful strategy or accurate future visions; rather, I implore you to experiment and explore these techniques that are available to construct and reconstruct your processes as needed, or as applicable to the client, goals, problem space, aspirations, or strategies you are setting out to discover. As long as you are able to stay organized, maintain a shared vocabulary and principles, and keep everyone informed about how and what you're investigating and what you are trying to accomplish, you will arrive at an aligned vision. Without a clear and aligned view of how you are futuring, the nonlinearity of Futures Thinking has the potential to confuse everyone or drive them mad, potentially derailing your goals, leadership, and aspirations. Some definitely prefer to use these methods in a particular order and are very strict about how and when to use them. But in talking to different practitioners, I've found that there are many ways to do futures work, and some concepts or methods are fundamental, while others are optional. It all depends on the practitioner, what they believe is effective, the project and goals, and who they are working with. Many are already creating new methods every day to supplement or replace traditional exercises, and it's an exciting time to see more and more people involved in advancing the practice to make it work for a variety of situations.

The advantage of futures work is that it is nonlinear and flexible for the purpose of deriving various perspectives about uncertainties. Inherently, it can be as malleable as you want it to be, and like any discipline, it's open to reinterpretation. As I've said before, there are no futures police officers or arbiters looking over your shoulder. There may be purists and academics who would die on their sword over how foresight should be practiced, but I can assure you that the world will not end if you use one method differently than it was intended or make up your own. In general, if you understand where you are going (problem framing) and are doing the appropriate research (analyzing and prioritizing trends and other futures intelligence) and using that information to form perspectives of the future (scenarios, implication mapping), it doesn't matter which methods you use, as long as your exploration is informed by what you discover, extrapolate, and generate so that you can form a clearer understanding of how the future could be. In Figure 8-24, I've outlined the various activities that can be used across the major stages of work. World building or scenario writing can be informed by any of the activities listed. Some of the output of exercises may differ in depth or perspective, so using multiple methods to understand future scenarios is always helpful for filling in the gaps.

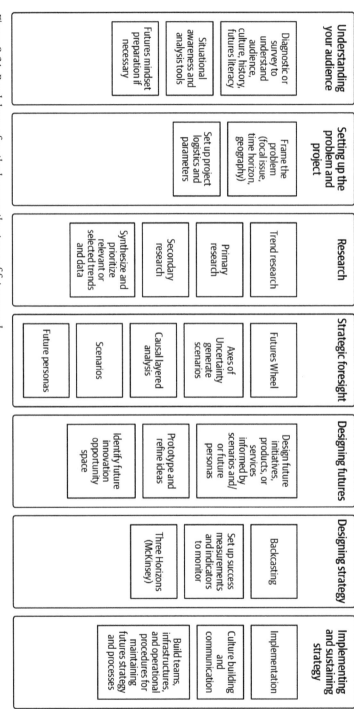

Figure 8-24. Breakdown of methods across the stages of futures work

That said, your audience could find this level of flexibility intimidating, paralyzing, or confusing. So be careful about how, when, and where you introduce, explain, or use these methods, and always make sure you have a method of prioritization (such as the Probability Versus Impact Matrix, quantified indexing,[9] or a voting mechanism[10]) to help everyone focus on what matters. Too many possibilities and options can be debilitating, and not enough options could feel shortsighted or superficial. Try to dial in what is "just enough" to deliver and inspire.

In Closing

Scenarios, uncertainty matrices, and implication maps are different ways of looking at "What if?" scenarios of the future. Many of these tools are essentially diagrams or maps toward and about the future. You might end up with several Futures Wheels, scenarios, or Axes of Uncertainties. Determine what is logical and useful for you to make sense of everything in a manner that feels safe and thorough for everyone on the team. And remember that just because you remove or ignore something during any stage of this process, that doesn't mean you are throwing anything away. Keep all your work in case you need to revisit it later. At times, you may feel like you are making confident decisions, but after exploring a particular future you realize it's not the "right" future. Going back to the drawing board to assess alternate views, paths, or options is absolutely OK. Futures work is nonlinear, but so is traversing the whole process. You can go forward and backward and laterally several times to see all the possibilities before you identify a single future or a subset of futures you want to align on.

---

9 Applying quantified data (market investment, projected value, the number of companies working in the sector/technology, the size of the industry) will allow you to put a weighted index to some of your data, saving you time in determining which trends, topics, or ideas you want to prioritize.

10 Voting is a subjective method of prioritizing and an absolutely valid method for selecting what matters the most to you or your team. Though some may think this method isn't empirical or data-driven, it can be a fast way of selecting important or relevant topics to pursue. Having stakeholders (leaders, finance, experts) in the room can also help establish that you aren't necessarily selecting based on biases.

As you wrap up foresight activities, try to think about the journey and how you've arrived at perspectives of potential futures. Think about how you will document the outcomes and priorities so that you can either communicate them out to stakeholders or use them for the next stages of work: designing futures and designing strategy. If you are using foresight only to map out possible future scenarios, your work could end here, but you might want to consider how you will visualize or enrich the meaning of those possibilities to make it accessible and inspiring to consume. Try to enhance your impact and reach by adding a richer visual, oral, or experiential dimension. After all the work you've done to reveal the future, why not find an effective way to engage your audience and ultimately incite discussion, change, or transformation? The next chapter will discuss how Design can play a role in bringing your scenarios and visions of the future to life through the use of imagination, creativity, prototyping, and experiential formats.

Checklist for Stage 1

1. Trends and Supplemental Research
   - You've rigorously collected key signals, drivers, and trends in the domain(s) you're exploring.
   - You've done primary and secondary research talking to experts and stakeholders and gathered information on current state and issues you see persisting into the future.

2. Research and Trend Analysis and Prioritization
   - You've categorized and tagged the trends that are relevant and important for your focal issue.
   - You've prioritized the trends and research that you want to move forward with.
   - You've mapped out multiple orders of implications of key trends, events, or signals.

3. Scenarios
   - You've had many conversations and collaborations to define the different types of futures that could arrive.
   - You've developed multiple scenarios using different methods and formats.
   - You have both positive and negative scenarios and various perspectives on possible future worlds.
   - You've extracted future personas from the scenarios or updated your initial personas or the population segments you're interested in.

Checklist for Stage 1

1. Trends and Supplemental Research
   - You've rigorously collected key signals, drivers, and trends in the domain(s) you're exploring.
   - You've done primary and secondary research talking to experts and stakeholders and gathered information on current state and issues you see persisting into the future.

2. Research and Trend Analysis and Prioritization
   - You've categorized and tagged the trends that are relevant and important for your focal issue.
   - You've prioritized the trends and research that you want to move forward with.
   - You've mapped out multiple orders of implications of key trends, events, or signals.

3. Scenarios
   - You've had many conversations and collaborations to define the different types of futures that could arrive.
   - You've developed multiple scenarios using different methods and formats.
   - You have both positive and negative scenarios and various perspectives on possible future worlds.
   - You've extracted future personas from the scenarios or updated your initial personas or the population segments you're interested in.

Designing Futures

In this stage, I'll discuss how Strategic Foresight informs design and innovation activities by providing the input for ideation and future vision concept development. Having been a practicing designer for over 20 years, and having been involved in innovation projects across many industries, I consider this to be the most exciting stage; it can also be the most challenging one if you are not well prepared. Designing a vision or an initiative for a distant future comes with its own risks when you are trying to determine what people might need but can't call the future. Thus I've collected many examples, methods, and approaches for how to make this stage more effective. That is why a whole chapter is dedicated to designing the ideation workshop. Preparing the environment, teams, and methods has always been important to me in my work as a facilitator, and when done effectively, this can be the most transformational and exciting stage of work.

In this stage, we'll discuss the following:

- Chapter 9, "Designing Futures"
  - Four common types of futures design work: Speculative Design, Design Fiction, Science Fiction Prototyping, and Experiential Futures
  - How futures design can be used as a prototyping tool, and how it can be effectively used as a strategic vision or North Star for an organization
- Chapter 10, "Designing the Ideation Workshop"
  - How to plan for a futures design workshop
  - Ideation methods, and how and when to use certain exercises depending on the constraints and the output you are looking for

Designing Futures

The significance of speculative design in strategic thinking and planning lies in its capacity to provide forward-thinking, holistic insights into the potential evolution of your product, service, or experience. Despite the uncertainty of the future, gaining a panoramic view and preparing for what lies ahead can be an invaluable asset for any company.

—Hiram Aragón, Design Lead, Products and Experiences, frog

Design can play a pivotal role in Futures Thinking. And while today the label "Design Thinking" is scrutinized, shunned, or even deemed extinct by some, we continue to use what we've learned from the advancement of this practice to imagine and create well into the future, no matter what we decide to call it. Traditionally, Futures Thinking, or Futures Studies, hasn't really formally included traditional design methods or processes as an integrated or essential function within its dialogue or approach. Not that Design has been entirely absent from futures work: people have been *designing* the future for a long time through various applications, channeling their imagination and ability to pick up on trends and cultural shifts around them.

Today, due to a growing global interest from the Strategic Foresight, Future Studies, and Design communities, we are seeing more designers who are adopting foresight and more futurists who are inviting designers into their work. But Design can play a more important role in the enrichment of futures and foresight work through its innate ability to assist in discovering and analyzing insights, uncovering opportunities, and synthesizing that information into a variety of novel visions, responses, or solutions. Design also uses rapid iterative capabilities to test and learn. And as a naturally creative function, it has the ability to harness the imagination while utilizing an arsenal of tools taken from systems thinking and service design, human computer interactions, human factors, architecture, engineering, anthropology, and social and behavioral sciences, as well as from many other scientific fields. Whether we are designing for the immediate tomorrow or for a far-off future, Design has many sensibilities that can enhance futures work when included and enabled throughout the entire process. Essentially, Design can:

- Provide additional ethnographic research and interview tactics and perspectives

- Provide additional methods for any stage of the process

- Synthesize research and ideas into new and novel responses

- Visualize and illustrate scenarios, products, and services

- Prototype ideas, environments, and behaviors

- Iterate quickly based on feedback or commentary

- Focus on details of interactions or systems design

- Incorporate and consider anthropology, psychology, and social and behavioral sciences

- Bring worlds, contexts, and personas to life through a variety of formats (2D, 3D, virtual, animation, or video/film)

This is not to say that other fields (such as architecture or engineering) are not capable of similar contributions! However, it's becoming apparent that Futures Thinking is not too dissimilar from what designers, researchers, and strategists are doing today. It's just an additional set of tools and perspectives. Not only does Design allow us to dream and transform those dreams into tangible realities, but it's also a means to address complex questions, fostering reflection and cocreation. It can embody an immersive experience or be an image or artifact so that one can actually *touch* or *experience* the future, respond to it, and give it a platform to engage in discourse.

In this chapter, we look at designing futures as a stage of processes of its own, one that can take the output from Strategic Foresight activities (trends, scenarios, implications) and use the information about the future world to develop innovative products, services, initiatives, or systems that ultimately culminate as your future vision or North Star—the future you want to build. Inversely, you can use this phase to develop the future you don't want as a provocation or a way to discuss threats or questions about what you don't want to happen. I'll break down some of the different genres, labels, and methods for designing future ideas and some of the nuances that make them similar or different. Ultimately, there are many ways to use Design in futures work, and I suggest you experiment with them (or make up your own) anywhere and everywhere you see fit for your project and strategy.

Speculative Design

My introduction into the world of futures began in 2009 with what was called *Critical Design*. Though Critical Design had its roots in a particular type of design for futures or alternate realities, as the practice grew in popularity and was critiqued, it eventually adopted the additional modification of *Speculative and Critical Design*. Through the decades, as speculation, critical theory, and design futures methods and communities evolved and emerged, *Speculative Design* began to be used more widely as an umbrella term to refer to many formats of design that "speculate" about potential future worlds, products, or services.

In an interview with Sara Božanić and Petra Bertalanič for the Speculative-Edu project (*https://oreil.ly/YEZJ8*), James Auger talked about the techniques used in creating convincing Speculative Design projects:

> *The crafting of a complex narrative or artifice using the real-life ecology where the fictitious concept is to be applied, and taking advantage of the nuances of contemporary media, familiar settings and complex human desires or fears. The careful blurring of fact and fiction changes the way the audience perceives the concept and in turn their reaction to it.*

You may hear terms such as *Design Fiction* and *Science Fiction Prototyping* (which I'll discuss later), among others; these practices have characteristics and methods that are similar or analogous to what Critical Design attempts to investigate, yet each has its nuances, its exemplars and pedagogy. Are they all speculative? Yes. Are they all a form of design? Surely. Can they all be discursive or provocative? Absolutely. Are they all related to the future? Not entirely, but mostly—especially if you think of the *future* as any moment beyond the present moment we are experiencing right now.

Each of these practices highlights the ways designers and futurists alike have tried to evolve the field of futuring through design across different contexts and purposes. All can be useful to some degree, depending on the application or desired impact. Speculative Design, among other versions of futures design, has the ability to challenge the status quo and our assumptions of what design could be now and in the future, but it also challenges why, how, and for whom we are designing, while traversing a parable that may or may not exist but certainly has the potential to.

Personally, I have been using *Speculative Design* as a label for any designed object, service, experience, or vision that represents something or some place in the future (or any time beyond the current moment). I use it as a way to describe many approaches for speculation and anticipation so that we can have a more simplified and accessible discussion about what speculation about the future means. In the last decade, this term has begun to free itself from its academic associations of purely discursive work, which has allowed it to operate as more than a thought experiment. This in turn allowed it to become more flexible and easily adopted. However, the evolution and acceptance of the speculative practice across industries is still in its infancy.

From what I've observed, there are more people who are aware of Speculative Design today than there were 10 years ago, but adoption is still moving slowly. We're not yet at a point where Speculative Design is on every designer's resume, but that is changing. Part of the stigma still attached to this label (and to other futures design genres) is its association with more conceptual and academic work, but as more of us create more evidence for how this format can lead to actual innovative ideas and to more responsible and holistic long-term strategies, the more credibility and value we can achieve across industries. However, regardless of what we call it or how we practice futures design, it is all *just design*.

In a 2015 article on *Medium* (*https://oreil.ly/wnsgR*), Cameron Tonkinwise, then a professor at Carnegie Mellon University, attempted to debate and reconcile all the different terms of design futuring that have emerged over the decades as variations of traditional design practice:

Every time you qualify design with, or add design to, some other quality or practice, you are claiming that design does not already do that. All these phrases are redundant and/or appropriative of design: Design Futures, Design Fiction, Speculative Design, Critical Design, Adversarial Design, Discursive Design, Interrogative Design, Design Probes, Ludic Design.[1] Designing that does not already Future, Fiction, Speculate, Criticize, Provoke, Discourse, Interrogate, Probe, Play, is inadequate designing. Speculations are forms of disciplined imagining, methods by which designers force themselves to think in more ambitiously counterfactual ways. Speculations try to push beyond current expectations and trending futures; they expand the sense of what is possible.

Many of the approaches mentioned in Tonkinwise's article embody characteristics of play, inquiry, speculation, or human/social/cultural investigation. And while they each have their own flavor as far as how they engage their audience and for what purpose, they are just variations of design research or prototyping. Tonkinwise's article also encapsulates much of the frustration I have observed watching practitioners who have tried to claim territories, trademark tools and processes, or debate the intent, formats, labels, or methodologies. Debates about ownership or nomenclature hold back the real work that is necessary for us to get the good collaborative work done. Once you are able to strip away the labels and dogma, they are all just a set of mechanics or vessels for creativity and imagination. It doesn't matter what you call your design approach as long as it is useful for progress and constructive debate. Design is all about breaking down the old, transforming the new, and reforming, reformatting, and evolving the world around us. So feel free to find your own path, create your own approaches, and help us all by evolving the practice. Start a dialogue and discuss what inspires others, and take caution with whatever triggers fears or creates obstacles for your work.

---

1 For information on Adversarial Design, Discursive Design, Interrogative Design, design probes, and Ludic Design, see the Glossary.

The practice of designing futures has evolved greatly over the last two decades. Just as the term *User Experience* was derived from what we used to refer to as Information Architecture, or Human-Computer Interaction, our practice will continue to evolve, specialize, splinter, and be renamed as society and the world continue to experiment and explore. If you are confused about which term you should be using, try to focus more on the methods and outcomes you want. Design, no matter what you call it or what time horizon you are designing for, is all just design. Just as *architecture* is still architecture, even though there are many forms and speculative approaches. Use these terms and approaches at your own discretion, and decide what works best in your situation.

Given these nuances between genres, the lines may seem blurred. So for the purpose of this book, I'll frame design futuring or futures designing into two categories: *design for provocation* (including discourse, critique, or debate) and *design for strategy* (an idea you want to build). How you use these methods is not strictly binary, and you will sometimes find that an idea will straddle both worlds or lead from provocation into strategy (or vice versa). But first let's dive into some of the different types and formats of designing for the future.

Types and Formats of Futures Design

In the beginning, I had very little understanding of *how* to create a Speculative Design. I discovered a few projects (mostly from universities like the Royal College of Art) and spent a lot of time trying to reverse engineer my favorite ones. One particular project by Anthony Dunne and Fiona Raby called *Technological Dreams Series: no 1, Robots* exhibited a collection of future "robots" (Figures 9-1 and 9-2). But Dunne and Raby's robots didn't look like the anthropomorphized robots I was used to seeing in sci-fi films and television shows. They were more like pieces of furniture that one might have to hold in a particular way to activate or interact with them. In their artists' statement for the project (*https://oreil.ly/xmIrp*), Dunne and Raby speculated that "robots are destined to play a significant part in our daily lives—not as super smart, functional machines, nor as pseudo life forms, but as technological cohabitants. But how will we interact with them? What new interdependencies and relationships might emerge in relation to different levels of robot intelligence and capability?" This project became a seminal example of how different a future could be if we thought about it from an alternative ethical, behavioral, or sociocultural perspective—one that was detached from what science fiction or other technocentric futures had popularized, and one that removed the fluff of drama and focused on other provocative dimensions in a world unlike today but with enough familiarity to pull you in through other philosophical lenses. This was a new type of design and a different kind of future visioning that I couldn't ignore. It became the core of my master's thesis to prove that this format of design could be useful to a traditional design practice. I had no visibility into their academic program, but I had many questions. What inspires this kind of work? What methods were used? What research was this based on? Why did it all feel so strange yet familiar? What was the emotional trigger I was feeling when viewing this work? And were there other formats of designing for the future?

Figure 9-1. Dunne and Raby, Technological Dreams Series: no 1, Robots, *MOMA (2007)*

Figure 9-2. Dunne and Raby, Technological Dreams Series: no 1, Robots, *MOMA (2007).*
From the artists' project statement: "Robot 3: This robot is a sentinel, it uses retinal scanning
technology to decide who accesses our data. In films iris scanning is always based on a quick
glance. This robot demands that you stare into its eyes for a long time, it needs to be sure it is
you. On another level, it asks what new forms of furniture might evolve in response to future
technological developments."

It wasn't until I discovered Strategic Foresight that I realized the similarities between the disciplines, and that there were many more examples and methods for how we might gather information about the future so that we could speculate on what future scenarios (and designed products or services) could look and feel like. It began to make a lot more sense to integrate foresight methods into the process instead of guessing what future we were going to create and why. Harnessing the power of trends and scenarios allowed us to anchor our design ideas into more specific worlds that were influenced by particular trends and conditions. It gave us more direction for research to respond to, and it allowed us to provide some statistical research that would serve as a basis for why we were proposing such ideas, what the product's implications were, and why we might want to build it (or not build it). Thus, FORESIGHT + DESIGN + STRATEGY became the logical formula I would bring into my own career as a designer and aspiring strategist. In my research, I discovered at least three methods of design futuring that seemed popular (some more than others), each offering unique and interesting ways of designing and prototyping future concepts: Critical Design, Design Fiction, and Science Fiction Prototyping.

CRITICAL DESIGN

Anthony Dunne coined the term *Critical Design* in the late 1990s in his book *Hertzian Tales* (MIT Press), where he referred to it as an attitude toward design as a critical investigation or lens into the possibilities of design in alternate realities or futures. Dunne wrote that Critical Design "is design as a mode of enquiry, using designed artifacts as an investigative medium to stimulate discussion and debate amongst designers, industry and the public about the social, cultural and ethical implications of existing and emerging technologies." Critical Design is best known for its use of *diegetic prototypes*, a term taken from theater and cinema in which props are used to depict objects, technologies, or other artifacts to support the narrative, environment, or realism of a story.[2] In futures they are typically "a technology that does not yet exist in the real world but is considered real and functional in the fictional narrative."[3] Such prototypes were useful in presenting viewers with a version of a world through the examination of a prop that lives in that world (as well as other props that could provide context to the conditions of that world). Much could be inferred by the design of a diegetic prototype, and when done well, it could fill in many ideas about what that world could be like, the actors who might use such artifacts, and what the implications could be were that type of object to exist in the world someday.

---

2 "'Diegetic prototypes'...account for the ways in which cinematic depictions of future technologies demonstrate to large public audiences a technology's need, viability and benevolence. Entertainment producers create diegetic prototypes by influencing dialogue, plot rationalizations, character interactions and narrative structure. These technologies only exist in the fictional world—what film scholars call the diegesis—but they exist as fully functioning objects in that world." From David Kirby, "The Future Is Now: Diegetic Prototypes and the Role of Popular Films in Generating Real-World Technological Development" (*https://oreil.ly/EFeYW*), *Social Studies of Science* 40, no. 1 (February 2010): 41–70.

3 Aloha Hufana Ambe, Margot Brereton, Alessandra Soro, Laurie Buys, and Paul Roe, "The Adventures of Older Authors: Exploring Futures Through Co-design Fictions," *CHI '19: Proceedings of the 2019 CHI Conference on Human Factors in Computing Systems* (New York: ACM, 2019).

Diegetic Prototypes in the Movie *Her*

In Spike Jonze's movie *Her*, the main character, Theodore Twombly, grows attached to a powerful AI operating system (OS). While this was a speculative and potential future, many diegetic prototypes were used to suggest how technology is embodied in this world. One of the devices Twombly uses to access the OS is a small book that looks like a cigarette box or picture album. It's a simple character and prop statement, but one that portrays how technology evolved and also hints at nostalgia and at what cultural trends or norms might exist in this world as well.

In a 2023 *Journal of Futures Studies* article (*https://oreil.ly/hU17B*) about the work of the film's production design team, led by K. K. Barrett, the writer Li Li noted that

> the possibilities of the future are not limited to fancy novel high-tech stuff; some nostalgic elements may also be an option. The device, according to Barrett, might have been made as a more sophisticated modern "thin sheet of curved glass," ...but instead, the Barrett team created a tactile, mundane, and real item, something similar to an old-fashioned cigarette box [Figure 9-3]. This nostalgic design is far from the fancy creation of the science fiction stereotypes. In contrast to the operating system's powerful ability to learn and create, the device's exterior design as one of the system's carriers appears outdated or nostalgic. The way we live today may be science fiction in our ancestors' eyes, but they may still prefer to live without the benefit of fancy technology. Similarly, modern people are frequently nostalgic for the good old days and tend to stick to their rituals. This diegetic prototype demonstrates that, in the near future, people may nostalgically cherish some poetic elements of their ancient ancestors while enjoying the comfort and convenience of advanced technology. The future does not have to be new. It could instead be a memorial to our ties to our former selves. Our current critical reflections on how modern technology affects us and what we prefer may lead to a very different future.

Figure 9-3. OS device in the movie Her *(source: Li Li, "Diegetic Prototypes in the Design Fiction Film Her: A Posthumanist Interpretation" (https://oreil.ly/hU17B/))*

Anthony Dunne and Fiona Raby, who led the Designing Interactions program at the Royal College of Art in London from 1990 to 2015, defined a generation of designers and design practice that would challenge the pedagogy of product design by offering a way to pose questions about possible futures and consequences and discuss alternative provocations for what our present-day choices mean for tomorrow. Highly criticized for its conceptual, inaccessible, or artistic nature as a commentary and its reluctance to focus on the actual needs or situations of people today, much of this work felt darkly satirical and at times uncomfortable, which in turn spawned questions about its validity as a legitimate design practice versus artistic expression. While there wasn't necessarily any intention for Critical Design to be used to create usable products today or in the future, it surely provoked long-term thinking and questions about society, culture, politics, technology, ethics, and other aspects that had sorely been missed in traditional product design or strategic approaches at the time. Within a world that was addicted to Design Thinking and Human-Centered Design, Critical Design brought a fresh perspective on what design and research meant for future societies, and on what approaches could be taken to deliver important questions about what could happen should we continue down certain paths (or not). Still, several works became striking examples of alternate futures that were grounded in real research. Dunne and Raby's *Foragers* project (Figure 9-4) hinged on a 2011 report (*https://oreil.ly/VAwpr*) by the UN Food and Agriculture Organization (FAO) in which it estimated that farmers will have to produce 70% more food by 2050 to meet the needs of the world's expected nine-billion-plus population.

In their artists' statement (*https://oreil.ly/4Fdzi*) for *Foragers*, Dunne and Raby explained the project's intentions:

> For this project Dunne & Raby look at evolutionary processes and molecular technologies and how we can take control. The assumption is that governments and industry together will not solve the problem and that groups of people will need to use available knowledge to build their own solutions, bottom-up....
>
> As such, a group of people take their fate into their own hands and start building DIY devices. They use synthetic biology to create "microbial stomach bacteria," along with electronic and mechanical devices, to maximize the nutritional value of the urban environment, making-up for any shortcomings in the commercially available but increasingly limited diet. These people are the new urban foragers.
>
> *Foragers* is about the contrast between bottom-up and top-down responses to a massive problem and the role played by technical and scientific knowledge. It builds on existing cultures currently working on the edges of society, who may initially appear extreme and specialist— guerrilla gardeners, garage biologists, freegan gleamers etc. By adapting and expanding these strategies, they become models to speculate on what might happen in the future.

Figure 9-4. Dunne & Raby, Foragers *(2009): Device for extracting nutritional elements from the environment*

DESIGN FICTION

A term first coined by science fiction author Bruce Sterling and later popularized by Julian Bleecker's 2009 paper "Design Fiction: A Short Essay on Design, Science, Fact and Fiction" (*https://oreil.ly/xmraE*) and Bleecker's design studio, Near Future Laboratory, Design Fiction has become a method of its own to visualize and prototype near-future concepts. Like Critical Design, it makes use of diegetic prototypes as artifacts and platforms for discussion. Design Fiction has rapidly grown in popularity over the last decade as a viable, accessible, and useful way to prototype the future. It doesn't adhere to a particular set of rules, methods, or principles other than to accelerate a discussion toward visualizing the potential existence or use of an idea. It does, however, accentuate the use of imagination as a key ingredient in its method of discovery and ideation, as well as an awareness of current trends and themes in society. Blending brands and remixing future tech with old tech or common artifacts displays a unique approach that shows how the future actually arrives. Sometimes it is something extremely new, and sometimes it's just a remix or an amalgamation of, or an amendment to, the mundane objects we already use today.

As defined by Julian Bleecker, cofounder of Near Future Laboratory and author of *The Manual of Design Fiction* (Near Future Laboratory), Design Fiction is

> *a mix of science fact, design, and science fiction. It combines the tools of research, storytelling, and speculation, with the material crafting of objects which don't exist now, but plausibly could in some version of the near future. For the purposes of discussion, debate, and extrapolation about the future, the object created is imbued with a quasi-narrative function, somewhat like a prop in a movie. It helps tell a story about where our technology and societies may be headed.*

Near Future Laboratory's inclusion of the mundaneness of futuristic ideas also brings a sense of realism that allows Design Fiction to be accessible and yet provocative enough to stir debate or discussion about potential realities that could be on the horizon. Many of the collective's proposals are grounded in fact and fiction and certainly are inspired by the world around us today, but they can sometimes also include a humorous or sarcastic spin that adds a bit of playfulness to the work. This delicate balance among what is, what could be, what was, and the silliness of human behavior fortifies the effectiveness of this approach as a form of speculation that doesn't take itself too seriously, but seriously enough to stimulate important discussions about potential futures. Today, many use Design Fiction almost synonymously with Speculative Design, and many flavors have emerged in contrast to the style of Near Future Laboratory's signature projects. Many tools have also been developed to stimulate creativity and brainstorming, and I'm sure we will continue to see more of them as these tools become more normalized in everyday design activities.

According to Near Future Laboratory, a good Design Fiction is:

- *Easily digestible*, and hence is the choice of mundane archetype formats (the manual, the infomercial, a grocery store coupon sheet, a YouTube review, etc.).

- *Facilitated specifically to capture diverse viewpoints* and perspectives in an organization, embracing friction and messiness.

- *Able to circulate easily*—beyond boardrooms and corporate culture, beyond the white cube of cultural centers and museums, beyond like-minded people (friends, designers, futures researchers)—and reach different audiences: partners, project stakeholders, users, citizens, the general public.

- *Debatable.* It should spark discussion and debate about the issues at stake in the area we want to address. The artifacts are just a means to an end— a provocation for debate, an alibi to start a conversation, an opportunity to step back, listen, and make more well-rounded and better-informed decisions.

In 2015, Near Future Laboratory, in collaboration with the Mobile Life Center and Boris Design Studio in Stockholm, organized a Design Fiction workshop to spark a conversation regarding the futures of connected things and the Internet of Things. In an era when the IKEA catalog was still an iconic printed brochure available at all stores to showcase the company's current products, and we were being bombarded by "smart" things, from smart fridges to smart homes, the team designed a catalog (Figures 9-5 and 9-6) that would showcase both a sign of the times and a collection of near-future products that IKEA could possibly offer. Near Future Laboratory offered this explanation on the project's website (*https://oreil.ly/thlAy*):

> We used the Ikea Catalog as a Design Fiction artifact for its compelling ways to represent normal, ordinary, everyday life in many parts of the world. [It] contains the routine furnishings of a normative everyday life. The result is a container of life's essentials and accessories which can

be extrapolated from today's normal into tomorrow's normal. Our Design Fiction Ikea catalog is a way to talk about a near future. It is not a specification, nor is it an aspiration or prediction. The work the catalog does—like all Design Fictions—is to encourage conversations about the kinds of near futures we'd prefer, even if that requires us to represent near futures we fear.

Figure 9-5. Near Future Laboratory, cover of "An Ikea Catalog from the Near Future" (September 2015) (https://oreil.ly/46aU2)

Figure 9-6. Near Future Laboratory, page from "An Ikea Catalog from the Near Future" (September 2015) (https://oreil.ly/46aU2) featuring KLIPPA Gardening GnomeDrone ("Home garden monitoring/harvesting/tending/watering drone keeps your garden alive, active and ebullient") and DELNING self-subscribing food storage & pantry ("De-clutter your home by placing all fresh, perishable, frozen and diffusion foodstuffs in a central, safe, encrypted, biometric security storage unit")

SCIENCE FICTION PROTOTYPING

Science Fiction Prototyping refers to the idea of using a science fiction narrative to describe and explore the implications of futuristic technologies and the social structures enabled by them. While it has very similar attributes to what we consider Design Fiction or Speculative Design, it has its own roots in science fiction narratives (film, television, comics, literature) as the inspiration for its format and ideas. Science Fiction or Sci-Fi (SF) Prototyping can harness an entire story or narrative as a platform for discovery and discussion. Unlike Critical Design and Design Fiction—both of which utilize storytelling, but generally through props or environments as a focal point of discussion—sci-fi prototypes can be completely on paper as text or storyboarded in the form of a comic or a series of illustrations.

In his book *Science Fiction Prototyping* (Morgan and Claypool), Brian David Johnson states that the goal of SF prototypes is

> *to start a conversation about technology and the future. Science fiction gives us a language so that we can have a conversation about the future. SF prototypes are tools to develop that language. The stories, movies and comics that we make can get researchers, designers, scientists, engineers, professors, politicians, philosophers and just everyday average people thinking about science in a new and creative way by using science fueled stories that capture our imaginations.*

SF Prototyping is unique in the sense that its focus is not necessarily a particular object or technology but rather the story around it. It utilizes traditional formats of science fiction, usually in a visual format, to think about the other aspects of the complex lives of the actors in that world—their drama, struggles, conflicts, and victories. By imagining the future in a format that we are all accustomed to consuming, it allows us to see more than just the future object; we see the environment, the use, the other actors (human or nonhuman), right down to the details about how something might be used or misused. SF Prototyping is more than just storytelling—it is a form of speculation that boosts the narrative, rather than the object, to the forefront. As a futures design method, it can be complementary to many of the other approaches and bring more life and detail to any proposal.

To quote again from Brian David Johnson's book *Science Fiction Prototyping*:

> *SF prototypes allow us to create multiple worlds and a wide variety of futures so that we may study and explore the intricacies of modern science. They are a powerful tool meant to enhance the traditional practices of research and design. The discoveries that we make with these prototypes can be used to question and explore current thinking on a level we have not approached in the past, namely using multiple futures and realities to test the implications and intricacies of theory. Additionally, the output of the science fiction prototype can feed information back into the science and technology development process, investigating and shaping how a user might encounter, explore and ultimately use that technology.*

In his book, Johnson explores numerous instances of SF Prototyping from his tenure as the chief futurist at Intel, a microchip manufacturer. He also delves into his examinations of well-known sci-fi films such as *I, Robot* and *War Games*, which he considers manifestations of sci-fi prototypes. In my interview with Johnson, he said this about SF Prototyping:

> *It gives people something to latch on to, it gives them a world. It literally shows them what the future could be like, but it's based on fact; you can then start unpacking that future and you can start talking about it, but it gives them that language to do that. There's a great quote from Kate O'Keeffe, former director of Cisco's Innovation Labs: "People just can't imagine 5 to 10 years out, so if you give them a sci-fi prototype, they can see the future, understand its implications, and then start figuring out what to do about it." And so in that way, the fiction becomes an incredible tool.*

To create a sci-fi prototype, Johnson uses a process similar to traditional narrative arc storytelling. Beginning with the science you want to explore, you build up the elements you might want to illustrate by considering how the science is introduced into the world or into a specific context and the implications or ramifications it might have on people, and then expanding on those implications in a story to flesh out potential scenarios for how it could affect our lives for better or worse. Drama or other narrative devices may be used to express the critical threats or opportunities as well as to dig deeper into the actual lives of those who might be affected. This converges into a five-step process to guide you in your plot development (Figure 9-7).

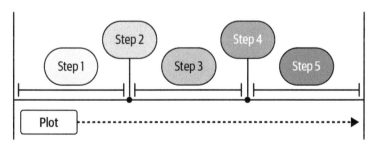

Figure 9-7. The five-step process of Science Fiction Prototyping (source: adapted from Brian David Johnson, Science Fiction Prototyping*)*

Step 1: Pick Your Science and Build Your World
First, pick the technology, science, or issue you want to explore with the prototype. Next, set up the world of your story and introduce us to the people and locations. You can answer very simple questions such as "Who are the main characters?" and "Where will the action take place?" You will also want to begin to explore an explanation of the technology in your topic.

Step 2: The Scientific Inflection Point
Introduce the "science" or technology you are looking to explore in the prototype.

Step 3: Ramifications of the Science on People
Explore the implications and ramifications of your science for the world you have described in Step 1. What effect does the technology have? How does it change people's lives? Does it create a new danger? What needs to be done to fix the problem?

Step 4: The Human Inflection Point
What did we learn from seeing the technology placed into a realistic setting? What needs to happen to fix the problem? Does the technology need to be modified? Is there a new area for experimentation or research?

Step 5: What Did We Learn?
Explore the possible implications, solution, or lessons learned from Step 4 and its human implications.

In 2008, Dr. Simon Egerton, Dr. Victor Callaghan, and Dr. Graham Clarke wrote a paper titled "Using Multiple Personas in Service Robots to Improve Exploration Strategies When Mapping New Environments" (*https:// oreil.ly/saZd7*). The paper explains the emerging research by the scientists, who were using a concept from philosophy and psychoanalysis in their robotics and AI development work. The paper also explores the benefits of building irrationality into the artificial intelligence of domestic robots to improve their ability to adapt to complex environments. The abstract of that paper includes this summary of their research:

> *In this paper we describe our work on geometrical and perception models for service robots that will support people living in future Digital Homes. We present a model that captures descriptions of both the physical and perceptual environment space. We present a summary of experimental*

results that show that these models are well suited to support service robot navigation in complex domestic worlds, such as digital homes. Finally, by way of introducing some of our current, but unpublished, research we present some ideas from philosophy and psychoanalytic studies which we use to speculate on the possibility of extending this model to include representations of persistent experiences in the form of multiple personas which we hypothesize might be applied to improve the performance of services robots by providing a mechanism to vary the balance of current and past experiences in control decisions which appear to serve people well.

Egerton and Johnson wrote a series of sci-fi stories, including one titled "Brain Machines" (Figure 9-8) that was based on a collection of scientific papers and hypotheses of the time. The story involved a robot named Jimmy, owned by none other than Egerton himself. The story is short and involves the simple task of Egerton asking Jimmy to fix him a gin and tonic. Throughout this task, Egerton has a fairly human conversation with Jimmy that in the background would require deeper AI logic and response mechanisms:

"Jimmy, fix me another drink," Dr. Simon Egerton said as he sat in his cramped apartment buried in the clog of stations that ringed the Earth.

"No problem," Jimmy replied cheerfully and set off for the make-shift bar.

"How are you feeling?" Egerton asked Jimmy.

"Fine thanks," he replied, mixing the gin and tonic. "We're running low on Tanqueray." He turned and waddled a few steps with the drink, concentrated and careful not to spill.

"Thanks." Egerton took the drink and searched the bot for anything out of the ordinary.

"No problem at all." Jimmy waddled back to the make-shift bar and tidied up.

Egerton sat his fresh drink on the floor next to the chair, lined up with eight other untouched cocktails. After a moment he asked, "Jimmy, will you fix me a drink?"

"No problem," Jimmy replied cheerfully and started on the tenth drink.

Egerton puzzled at the back of the little bot. He knew that Jimmy knew he wasn't drinking the gins. Egerton knew that Jimmy knew he was being tested and that it was silly to keep on making gin after gin. But Jimmy wouldn't react. He wouldn't break out of his service duties and ask what was going on. Why wasn't Egerton drinking the cocktails? Was there something wrong with them? It was a problem of will: free will. Jimmy had all the capabilities to question what was going on but he wouldn't do it. It was a problem Egerton had been trying to crack for over 6 months.

Figure 9-8. Illustration from "Brain Machines" by Brian David Johnson (source: Winkstink)

This work explores the role of multiple personalities in an artificial intelligence (AI) and both the positive and negative effects of instability and

irrationality on the system. The scene, where Jimmy is challenged to make multiple gin and tonics, provided the scientists with scenarios that they could build and test. Thus, the "Gin and Tonic Test" was born. The "Gin and Tonic Test" synthesized the various components of the AI theory, unifying them into a single scenario. **Brain Machines** *is also being used in an AI design competition from the Creative Science Foundation (https://oreil.ly/EBVwT). The competition challenges university students to develop the AI software needed to build Jimmy. As a part of the com-petition, the students are given access not only to the scientific papers and software needed for the experiment but they are also given the SF prototype* **Brain Machines** *to help inform their development. The first step in the competition is to build the AI with multiple personas to be tested in a virtual environment. The winners will then have their "Jimmys" uploaded to an actual robot [Figure 9-9] for further testing.*

Figure 9-9. Jimmy the Robot model (photo by Brian David Johnson)

Lowe's, a large retail home improvement company in the US, boasts innovation labs in which the company has been utilizing science fiction storytelling (prototyping) as the basis for its futures and innovation work. These "Narrative-Driven Innovation" principles are used to catalyze long-term visions and build actionable innovation roadmaps through emotionally resonant human-centered stories. In its labs, which are equipped with the latest technology, such as VR, AR, and mixed reality simulation environments, the company makes use of both real-world tech and science fiction storytelling. It hires professional science fiction writers and illustrators to develop stories about the future to serve as an innovation bed for new ideas. A simple comic strip (Figure 9-10) about a natural disaster and future-tech-enabled heroes eventually led to the discovery and development of new products such as in-store robots, exoskeletons for Lowe's staff, digital twins, and in-store visualizations (Figure 9-11) and simulations for its customers.

Figure 9-10. Science fiction comic from Lowe's Innovation Labs (https://oreil.ly/YvY2a)

Figure 9-11. Lowe's Innovation Labs, in-store visualizations (https://oreil.ly/zoUZQ)

The labs use Sci-Fi Prototyping as a storytelling device to discuss challenges in the future and today. But ultimately it is a combination of storytelling, imagination, emerging technology, physical and virtual prototyping, and user research that allows their process to lead to real-world applications and new intellectual property. The comics they write are just a starting point for a broader discussion about the needs of Lowe's employees and customers.

Neo Fruit

In 2019, Meydan Levy, an industrial designer and a graduate of Bezalel Academy of Art and Design, created *Neo Fruit* (Figures 9-12 and 9-13) for his final speculative technology project. *Neo Fruit* investigates humanity's relationship to food and today's challenges of food security, increasing populations, industrial food processing, distribution at scale, and an evolving focus on nutrition and diet. While the project doesn't necessarily state when, where, or which future Levy is focused on, it does suggest methods for manufacturing, synthesizing, and fabricating a new type of fruit for consumption by highlighting both the psychological and sensorial aspects of food consumption in its form. He developed a "4D printing" system that prints exotic and erotic silicone vessels for fruit ingredients to be injected into. This new kind of user experience creates an interaction that is sensual, fun, and alien-like. As an industrial designer, Levy was interested not only in the growing needs of future food systems but also in the production capabilities at scale that might need to be developed to engineer new types of foods for a growing population. As a potential marketing tool, his accompanying video (*https://oreil.ly/iCywG*), colorful and contemporary, brings the product into the average consumer's life so that it feels accessible and normalized, rather than seeming like a distant dystopian narrative in which the product becomes a necessity for survival. Levy's take on speculative design is as successful as it can be while also investigating the viability and feasibility of a new type of manufacturing system that would be necessary to produce this new food format. In the end, the project is in fact *speculative* because it does not exist yet, but it could be created today, and he is actively researching what it might take to bring this into reality as a startup. As a speculative proposal, he has found a way to harness the principles of anticipatory thinking, prototype it, and actually find a strategic path forward to make it a reality.

In an interview with *Core77* (*https://oreil.ly/6Dqf1*), Meydan Levy explained *Neo Fruit*:

> *This project offers a fresh artificial alternative to familiar fruits, conventional methods and technologies of production by putting together all of man's greatest capabilities and accomplishments to date, with nature's everlasting wisdom. My project offers a future optimistic way that combines technology with the emotional needs of human beings while taking responsibility for the environment in which we live. My fruit is a tool for perception changing and emphasizing the values that disappear from our lives that ultimately are responsible for our happiness and leave the "obvious" to wonder.*

Figure 9-12. Neo Fruit, *Meydan Levy*

Figure 9-13. Neo Fruit, *Meydan Levy*

Speculative Design for Discourse and Strategy

Now that we've looked at a few of the different formats and approaches to futures design, you might see many similarities and differences between how each is used for research, prototyping, provocation, or strategic initiatives. Those of you who are discovering these examples for the first time may be very confused about what label to give each one, how to practice it, or which is the right type to use for your projects in practical applications. I myself have gotten confused at times, and I try to avoid having debates around what characteristics make an example a Critical Design work versus a Design Fiction versus a general futuristic concept design. My advice is to not get too mired in terminology but to use the examples as inspiration and determine the appropriate terminology for your audience. Ironically, I'm sure these divisions will spur discourse and debate amongst many, but again, this is meant to provide an alternative framework for understanding the different methods and how we can make use of them practically for design and futures work today.

USING SPECULATIVE DESIGN FOR PROVOCATION, DISCOURSE, CRITIQUE, OR DEBATE

The intent of this type of futures design is to provoke or stimulate its audience to discuss the future scenario, ecosystem, product, service, or general idea. The intent may not necessarily be to make the idea into a multimillion-dollar revenue-building reality, but rather to bring forth important discussions about the potential of the idea, whether it should exist, or whether there are important aspects about technology, ethics, policy, behaviors, society, or other implications that the idea might generate. There may be zero intent for the idea to become a reality. Other types of design that are meant to inspire conversation or debate include Discursive Design, Interrogative Design, Adversarial Design, Design for Debate, Reflective Design, and Contestational Design. Speculative Design for discourse or provocation can be something that is purely relegated to academia or to cultural institutions such as museums. And while this type of design may seem to function only as a conversation piece, the work still provides very valuable and critical perceptions about the future and speaks to current themes, concerns, and aspirations that our current era, mindset, or society is thinking about and wants to expose to an audience for feedback or debate, or to activate new conversations about *what if, what could be,* or *what should* or *shouldn't be* in the future. It may not be the idea you want to build, but it may be used as a testament or statement toward a world you want to build. The "actionable" strategy correlated to this speculation could well develop or spawn from it later.

Unfortunately, organizations don't always want or need a conversation piece (whether or not it is a relevant and important one). Organizations, and especially profit-driven ones, typically will want to realize how they can make money or survive with this process. In a perfect world, we would have both efforts—*design for provocation,* which could stimulate important discussions, and *design for delivery (for strategy)* to invest in the future that we want tied to a clear strategy and roadmap.

But sometimes we have to make compromises and tune our sensibilities to providing something that is useful and intentional, and that our company or team truly wants to invest in. That investment could be a change based on discourse, or it could become the thing you actually want to create. I've worked with clients in many situations in which we push our thinking to imagine a seemingly impossible yet aspirational future (e.g., a quantum computer in everyone's pocket), we discuss its potential and its absurdities to unpack if/how/why we would want such a future product (what could we do if we had that level of power running AI services for us), and then perhaps we begin scaling back features to something that would seem either more reasonable or more attainable in the near future (quantum isn't ready for consumer applications yet, so let's choose a different technology that we know is available in the next three years), while still considering potential implications (what if quantum *does* accelerate into consumers' hands soon).

USING SPECULATIVE DESIGN FOR STRATEGY

The intent of this type of futures design is to deliver a preferable and plausible vision or concept about the future that we can realistically strategize and build toward. It is a speculation that has clear intentions to become real. Debate and discourse is surely part of the conversation and can help to refine the design concept even more, but the intention for creating the design is to develop a "North Star," a realistic aspirational vision of the future product or service you want to build toward. I say "build toward" because it is just a guiding star at the moment—it may have a tendency to change and transform as you test and prototype components of it today and find out more about what is right or wrong about your research or speculation or what needs to shift based on changing patterns in the world.

It's important to accept that this vision can change over time (failure, misunderstanding, or inaccuracies may exist!) due to the constant variability of future events as they arrive. You might say, "Well, that's just traditional design strategy"—and you would be exactly right! The only difference is that we are employing additional or alternative tools to get there. Futures can provide us with more holistic views of future conditions. But everything beyond this moment is a speculation—and this is why "speculative" design surely has a place in traditional design strategy as a tool and framework.

These are not by any means the only categories or ways in which design can be used within Futures Thinking. There do exist some similar concept-testing formats that are not specifically called Speculative Design, such as Concept Cars,[4]

---

[4] Typically associated with the automotive industry to represent future concept visions of automobiles that are displayed at car shows, the term *Concept Cars* is also used to refer to any kind of future design in other industries (digital, industrial design). In a paper presented at the 23rd Innovation and Product Development Management Conference in 2016 (*https://oreil.ly/YPnjC*), Ricardo Mejia Sarmiento and his coauthors describe Concept Cars as offering "a design-led approach of researching the future, where visual synthesis, prototyping, and storytelling play an important role. Concept Cars act as probes that simultaneously explore technologies and styling while also communicating a probable, plausible, and preferable future, in one time-horizon. Unlike managerial futures techniques, Concept Cars provide tangible futures that people with different backgrounds can experience, influencing several parties involved in developing an innovation."

Value Proposition Testing,[5] or Sacrificial Concepts.[6] A provocative debate can certainly lead to the development of a strategic roadmap. Inversely, a strategic vision could engender a provocative debate about certain topics or features you've proposed. There is no right or wrong or prescriptive sequence for how, when, or where to use these tools, as long as you are able to use them to effectively envision the future you want or don't want. How you enable the tools and what you do with them is up to you and your team or organization. Using the various types of design approaches will allow you to focus your intentions and outputs so that they can bring value to your practice, organization, or business. When we think about what it takes to be pioneering and innovative, this is the phase where we can flex that muscle.

Returning to our sailboat analogy, imagine that we are on the boat and sailing into that uncharted territory. We've had a lot of discussions about what the island might look like and feel like, what the weather conditions might be, and what animals and plants might be there. We will have to speculate on how we will survive there. We might create some small-scale model of the island to figure out where we will dock our ship, how we will gather food, and where we will set up camp. All of this is speculative in terms of what we want to build and how we will operate. We could build some small models of our camp and even do some light prototyping before we set sail. As we might learn more about the island over time, our vision of what we build there could change. Before embarking on our journey, we may even want to prototype what it's like to live together.

Colonizing another planet is an excellent platform for speculative design due to the lack of information, or the information that has to be inferred from the data we are able to gather. However, living on a planet like Mars comes with challenges that we don't yet even know exist. It requires speculation and prototyping here on earth before we can invest the millions of dollars to get there. In 2019, NASA's 3D printed habitat challenge saw a range of entries that

---

5 "Value propositions are the north stars that guide your business. When you get them right, they can be key business drivers and help you reach and monetize your target customers....[A value proposition] helps the entire organization, whether you're in accounting or engineering or product, really align on what it is you're trying to offer in the marketplace and move in the same direction." From "How to Test Value Propositions Like a Business Designer" (*https://oreil.ly/ijyLs*), IDEO U (n.d.).

6 According to a post by Natalie Vanns (*https://oreil.ly/zeGRD*) on the website of the design agency Mace and Menter, sacrificial concepts "are a tool originally developed by the design firm IDEO. They are used in early research as a stimulus for discussion, and are different from presenting prototypes later in the design process for the purposes of testing or validation."

envisioned how we might construct a habitat on Mars when the time comes to populate the planet. AI Space Factory, an architectural and engineering firm whose mission is to enable human habitation beyond Earth, won the contest with its Marsha concept (Figures 9-14 and 9-15), which utilized 3D printing. The printer would use the Martian soil, converting it into a biopolymer basalt composite—a biodegradable and recyclable material derived from natural materials found on Mars.

Figure 9-14. Exterior of Marsha Space Habitat, AI Space Factory

Figure 9-15. Interior of Marsha Space Habitat, AI Space Factory

Architects have been using speculative thinking and design for many years prior to the advent of Speculative Design, and while they may not use a particular methodology such as Strategic Foresight, they still use very similar lenses when investigating future trends, materials, environmental impact, engineering, and human behaviors. Due to the complexity of architecture and its safety regulations, it's important for architects to consider any and every implication that could put the integrity of a project at risk. Since the production cycle of architectural projects can also take many years, even decades, it is imperative that they think as far out as possible, not only to make sure their work can be realized but also to ensure that it can withstand sometimes volatile and unpredictable environments. Similarly, they are interested in VUCA and anything else that could compromise their work in the near or far future.

Prototyping the Future Today

Engaging in the design phase means you are intentionally using the craft of design to develop an idea into something that can be experienced beyond just words on a page. We can say that with design we are prototyping the future. We can do this in many ways, but essentially it's a way for us to respond with a variety of lenses to decide if this is the future we want or don't want. Prototyping should not be new to designers and is in fact one of design's key tenets for creating desirable products. Futures can employ similar techniques, using the principles of design fiction or diegetic prototypes to garner feedback or reactions from people today. A prototype can be used as the artifact that you wish to create a conversation around, or it can be the actual vision of the future product or service.

The good news about futures is that whatever idea you come up with doesn't have to stay in the future. Sure, you may have projected it as something that society will eventually need in 5 or 10 years, but you don't have to wait to start building it. While there are some conditions or technologies that might not be available yet, you can begin to test the idea early to work out any bugs or fallacies. You can begin building part of the idea that is possible to manufacture today. Prototyping your idea now could in turn deliver the first phase or version of the grander vision. And who knows, testing the idea in the world today could stimulate others to begin helping, thus accelerating the technology or other conditions to make the grander vision a reality even sooner. That is the beauty of using futures as an innovation process. We are basically looking ahead so that we can determine what might be needed today to create the future we want. Then we just create it.

> *Prototypes serve as catalysts for sparking conversations, eliciting reactions, and evoking sensations that ignite the imagination. In my view, this is the essence of futures design. As designers, our role is to champion all forms of prototyping, provided they help us tackle present or future challenges in innovative ways.*
>
> —Hiram Aragón, Design Lead, Products and Experiences, frog

If you have the opportunity and budget to construct physical and interactive prototypes of your future design idea, that can be a powerful tool for feedback (whether for discourse or strategy). An artifact or experience that we use as a

prop or representation of a future design concept can either be discursive or be used to manifest an idea into a physical or immersive form so that an audience can touch, feel, and experience that idea to provide feedback for refinement, usability, or preferability. Prototypes are a great way of showing people what the future looks and feels like so that they can understand what it's like to use such a product or service and decide if it is the concept they want to create or avoid.

Futurity Systems, a futures consultancy based in Barcelona, Spain, is an invention laboratory and design firm that uses science and technology as its core ability to imagine futures. The firm's team of designers and technologists create working physical (and virtual) prototypes of concepts they conceive so that they can show their clients the potential of future technology, gain buy-in to hire their services, and use the prototypes as a launching point for conversations and other innovation efforts. One project that they use regularly is called the Electronic Popsicle (Figure 9-16); it is a device that you place on your tongue and that is connected to copper electrodes that target taste sensors on the tongue and can activate different taste sensations through electrical currents that are programmed to stimulate a combination of taste sensations on the tongue's surface. By stimulating sweet, salty, or bitter taste sensations, they can create the illusion of tasting different flavors without actual food being involved.

In an interview with Cecilia Tham and Mark Bünger, cofounders of Futurity Systems, Cecilia explains her invention:

> With the Electronic Popsicle, we use a very simple technology. But it's not even an advanced transformative technology, it's literally just electricity, your tongue, and your brain, but we are applying it in a very different way. How does this product affect the way we taste, and can you carry that application further into discussing different business models? You can begin to induce taste sensations without the presence of actual food. What if we combined that with an amplifier and connected it to music so that you could literally taste music? How could Spotify's services change in the future? Once we carry our clients through this exercise, they are inspired to imagine what kind of new products and services are possible in the future. When you have a prototype, you have a story to tell, you have a prop. When you show this to a client, they now have something they can show their colleagues—either as an aspirational example, or as evidence that work has been done.

The Electronic Popsicle is just one of many inventions Futurity Systems has conceived to test and highlight emerging technologies in different contexts. Some of the inventions come directly from the minds of its founders and employees, but they are always inspired by scientific rigor. Among the many tools they use to understand trends and emerging fields is a platform they are developing that utilizes a complex set of algorithms to search both public and private sources for information on scientific developments or trending topics. The platform allows them to keep an eye on real-time information as it evolves across the world and stimulate ideas about what people are working on, what people are interested in, and where innovation opportunities might be possible.

Figure 9-16. The Electronic Popsicle, Futurity Systems

Futurity also prototypes within virtual environments using VR and AR technology. Anticipating the metaverse, it developed the Store of Future Things (SOFT), where you can browse various galleries of nonfungible tokens (NFTs; see Figure 9-17). Within the environment you can browse galleries in different formats—a temple, a living room, a platform with no walls. By experimenting with the virtual space necessary to view virtual artwork, Futurity tests the literal boundaries of the experience and what might be effective and necessary for a future retail or cultural experience in the metaverse.

Figure 9-17. The Store of Future Things (SOFT), Futurity Systems

You can also experience the laboratory's first virtual plant ecosystem, which Futurity calls the *Plantiverse*; it includes virtual data visualizations of plants living in the real world that are connected to a variety of sensors. The sensors feed data into digital plants in the virtual environment, displaying everything from the amount of sunlight they are getting to the mineral health of each plant. The virtual plants, which Futurity calls *NFTrees*, are an investigation into how we might begin to include organic life forms in the metaverse, what role and value they might have for science and nature, and the questions that we may want to answer about what exactly we need and want in an eventual metaverse to make it feel comfortable, safe, or analogous to the real world. As a virtual prototype, Futurity's SOFT experience allows it to quickly build out ideas and test scenarios so that it can prototype new ideas and products for its clients in a working interactive environment.

Personal Carbon Economy

Shihan Zhang, Design Futurist at Alter+ (Alter Plus), a studio envisioning preferable long-term futures, moved to San Francisco from Shanghai, China, a city with more than 26 million people as of 2019. According to a report from July 2014 (*https://oreil.ly/tRxRy*), at least 50% of Shanghai's air pollution is from vehicle and factory emissions. Along with Beijing, Shanghai boasts some of the worst air pollution in China. Growing up in a world in which air quality was an ongoing health risk, Shihan focused some of her initial speculative design proposals on the *Future of Breathing* (Figures 9-18 and 9-19) and how brands might engage in new products to help citizens cope with challenges of air pollution. For her final project, she created the *Personal Carbon Economy* (*https://oreil.ly/ZM3dm*), a suite of products focused on recapturing carbon from the atmosphere (Figures 9-20 and 9-21). Her question: what if carbon dioxide can be marketized and monetized on an individual scale? Her answer was a possible economic model and a variety of inventions (speculative products) offering a lens into a not-so-far-fetched near future of personal carbon cap and trade.

> *Something that I always do is to think about mundane artifacts to prototype for the futures so that people can resonate with them. Especially when I am working on environmental challenges. It always felt so distant and far in the future when talking about those topics, and audiences might find it to be a mental challenge to relate to and how it might impact them personally. So by creating artifacts that are inspired by everyday objects, it clicked, like, "Oh, that actually impacted me," and it allowed them to quickly illustrate the futures we presented.*
>
> —Shihan Zhang, Design Futurist at Alter+

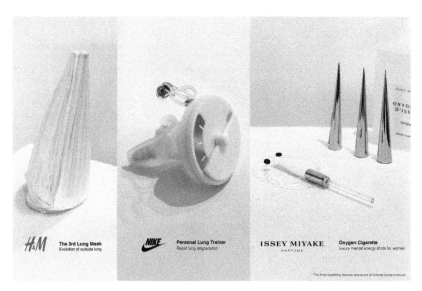

Figure 9-18. Speculative products from Shihan Zhang's The Future of Breathing: *the 3rd Lung Mask by H&M, the Personal Lung Trainer by Nike, and the Oxygen Cigarette by Issey Miyake*

Figure 9-19. The Future of Breathing, *Shihan Zhang: an advertisement for one of Shihan's speculative products, Oxygen Cigarettes by Issey Miyake*

Figure 9-20. Personal Carbon Economy (https://oreil.ly/B5NTv), SKIN Farming jacket, Shihan Zhang: the jacket uses body heat to cultivate algae and offset CO_2; wearing it is envisioned as a future side job for extra income

Figure 9-21. Personal Carbon Economy (https://oreil.ly/B5NTv), SKIN Garden dress, Shihan Zhang: carbon sustainability has become a value system rather than a purely political behavior within the Carbon Union. Citizens are prompted to embrace new symbiotic lifestyles with algae to reduce their carbon emissions.

Shihan was also very interested in how brands could bring future products into the world and how society and culture attach to brands, identities, and the different styles and values that each brand could bring to design. Thus bringing Speculative Design into a large tech organization like Twitch, where she works as a Design Lead today, had become a mission for her as she continued to practice with her studio, Alter+, outside of her day job. She experimented with many ideas for how to introduce the process of collecting signals and trends, testing methods, and finding practical applications that both the design teams and leadership can use and support.

Shihan explained her process of collecting signals, trends, and inspiration:

> What works great for our studio is the practice of collecting signals in our daily life, which might not feel helpful in the moment when you are doing it but can provide value and inspiration later. I'm always on LinkedIn or Instagram or other social media platforms. If I see something that looks interesting, I just take a screenshot and put it into a folder. Some signal stays in my brain even though I don't remember the details. Whenever there's a project I need to work on, my brain might recall that signal. If there's something specific I remember I would just dig into the folder and be ready to get inspired. Sometimes I'm also able to make connections across multiple signals and spark new project directions.

In Closing

Design can play a critical role in visualizing and materializing future visions. As we've seen already, there can be a lot of research, analysis, and extrapolation involved in curating what potential future scenarios and environments can look like. Using Design as an exploratory and innovation tool to transport you into that future can be very powerful in crafting visionary products, services, and systems we want (or don't want). But design is not a magical talent endowed by the gods; it's an art and a science that harnesses experience and the imagination inspired by research, the world around us, and our dreams, and it must be facilitated. No matter who you are working with, it can be challenging to generate something that is truly inspiring and isn't just some regurgitation of old ideas that might look like your competition. That's why I've dedicated the next chapter to designing the ideation workshop experience. These visions can and will ultimately become your North Star. By allocating a proper amount of time, resources, and activities to this phase, you have the ability to stimulate more creativity and generate more novel ideas. Not to mention it can also be quite fun to brainstorm and collaborate to articulate dreams and desires. Remember to deliver a future vision people want to invest in; it needs to be something powerful and provocative enough for them to want that future to materialize. Your vision needs to have a strong message, with strong roots in your research, implications, and scenarios, and it must have the potential to create a preferable future we want or possibly avert us from the futures we don't want. In the next chapter, I'll discuss how to set up the workshop and a few useful methods, mechanics, and use cases to provide you with a range of tools and formats for designing the future.

Designing the Ideation Workshop

Once you reach this stage of designing a future product or service, you may want to have a pulse check with your client and run another diagnostic to get a sense of their investment, excitement, and readiness for moving into this phase of brainstorming ideas. If for some reason they are not clear on what you will be doing, you can use some of the same tactics you used at the beginning of this journey to prepare (prime) them for the futures mindset, but this time with a focus on getting them into a DESIGN-oriented mindset—a way of thinking and participating that engages imagination and creativity to generate ideas and tangible visions. In this chapter I'll discuss various ways to prepare participants, as well as some popular methods and tools you can use for ideation and brainstorming.

Preparing for the Workshop

Preparing people to *create* can be an entire preamble phase within itself and can involve tactics similar to those that were discussed in Stage 0, such as showing them examples of outputs, revealing some of the methods and how they work, getting them comfortable with creating low-fidelity sketching and prototyping, and generally creating a safe space for them to imagine while channeling everything they've learned during the foresight process. Sometimes simply calling your ideation workshop "The Future of [insert company name]" or "The Future of AI in [insert company name]" gives participants a sense of privilege and investment, and the feeling that they are being invited to a very important meeting that could decide the fate and strategy of the company's future. Making sure people feel like they are integral to this process breaks down any stigma that this is just another meeting about current state problems and signals instead that this is an important event that requires their expert opinions, inputs, and aspirations.

Prior to beginning any design ideation activity, you'll want to make sure you prepare yourself and all the participants for the activity. This is a critical moment in which you are planning to explore more detail about a future product, service, or scenario. Jumping too hastily into ideation could pose risks to the quality of the output and threaten the overall outcome. After all the time you may have spent in the foresight stages weeding through signals, trends, and scenarios, you'll want to give yourself adequate time and a safe platform for people to explore and generate the most provocative and innovative ideas possible. Think about who you want to invite into this event—who is necessary for leadership, facilitation, craft, or alignment? It will depend on your team and resources, but the participants should likely include a facilitator(s), designers, illustrators, strategists, prototypers, executive leaders, the futures team, and domain or subject matter experts.

Let's look at that roster of suggested invitees in more detail:

Facilitator(s)

Every workshop needs a facilitator (*at least* one, if not more). Find a facilitator who has experience in conducting creative brainstorming workshops or knows the methods. They don't necessarily have to be a futures expert, but it's definitely a benefit if they know the process and the inputs (the outputs from the foresight and research) so that they can lean in and reference information to inspire ideas if necessary.

Designers

Depending on your project, team, and resources, designers (UX, UI, service, industrial, strategists, research) should be a part of this workshop because you are literally designing the future. Leverage their capabilities—their imagination, curiosity, craft, and desire—to create something novel, inspiring, and informed by the research and conditions you give them. Generating ideas from multiple perspectives, reframing problems and solutions, and pushing the boundaries and constraints are part of the everyday language they love to speak.

Illustrators

People who know how to draw or use generative artificial intelligence (AI) or who can sketch quickly are always an added benefit to any creative brainstorming workshop. Whether you are using them to sketchnote (a

method of taking notes visually using hand-drawn illustrations) or to illustrate people's ideas, they can definitely help to polish outputs.

Prototypers

Whether it's digital or physical prototyping, the ability to craft something quickly so that people can interact with it for feedback, commentary, and debate can make a big difference in the quality of the idea. While a 2D image or drawing is also powerful, if someone can quickly create a digital app or physically construct an artifact, you'll have the ability to touch, feel, and interact with it in a way that you can't always do with 2D images.

Executive leaders

Think about which leaders might need to be in the room to participate, collaborate, and observe and are necessary for making important decisions or aligning the teams. However, take caution when inviting executive leaders whose presence could intimidate participants, making it more difficult for them to think freely and creatively. Some leaders may also try to push their own ideas instead of allowing people to collaborate democratically.

Futures team

Anyone who's been involved in the futures project from the beginning should be involved in the workshop if they're available. They'll have the deepest knowledge of the research and scenarios, knowledge that will be needed to fuel the ideation and innovation process. They can be your subject matter experts and your designers and have already invested their time into getting you to this point.

Domain or subject matter experts (SMEs)

If you have access to any SMEs or domain experts, invite them. They can provide deeper knowledge and experience that can be valuable during the creative process. If they were involved in the foresight and research stages, having them as partners for brainstorming will be even more useful.

Looking back at what you've done in the Strategic Foresight stages, you'll want to determine what scenarios, implications, or trends you want to bring into the design workshop. Due to the nonlinearity of futures work, there can be several points of departure when ideating. If you haven't been prioritizing and converging on a smaller set of information about the future world or worlds along the way, you could easily become overwhelmed by the possibilities. Jake Dunagan, director of the Governance Futures Lab at Institute for the Future

(IFTF), recalls a concept and phrase coined by his colleague Stuart Candy—the "weight of alternatives":

> To make an analogy, one of the things that has come up a lot in our [referring to his work with Stuart Candy] conversations in situations like, say you're waiting on a job offer or moving somewhere or seeing if immigration papers are gonna go through—there's many uncertainties and many possible futures to assess and speculate upon. We started calling this the Weight of Alternatives. It's great to explore alternative futures, but to hold all of it together can be mentally difficult; it can be cognitively and emotionally hard to hold multiple futures in your mind while you are trying to design. So there's a trade-off. We don't want people thinking in singular linear futures. We're not doing any good by doing that, because that's typically our default setting anyway, so we're not adding value to that conversation. So we really want to push plurality. How can we make people see alternative futures and that there's not one path forward? It can be really hard, and it weighs people down when you have too many moving parts. It breaks. People become overloaded. So it helps to lighten the cognitive load. Where can I get the biggest bang for the buck? What are shortcuts I can take in the design process or the experience? How can I get people to that thinking without overloading them mentally and cognitively?

This weight or burden of generating many possibilities or options for the future can be quite debilitating when you are trying to converge on a particular set of problems or implications that you want to design a vision or solution for. This is why I always recommend having a prioritization method or activity in your process. If you aren't constantly making collaborative decisions on what to focus on, you end up carrying the weight of all your work with you until the end, making it virtually impossible (or at least very time consuming) to converge on a set of ideas or visions that you can effectively strategize and deliver on.

If you are shooting for developing an innovative product or service, then much like you tend to do in traditional product or service design, you'll want to align around a specific set of problems and/or people (or nonhuman actors) to solve for. To do this, you'll need to extract from your synthesized and prioritized foresight and research work to nail down what prospective situation, context, behavior, scenario, implication, persona, or environment you're interested in. What is the need and the dream you want to solve for, and how will your idea (though speculative at this point) address those issues? With the proper amount of input, you can generate a number of ideas that could solve for both collapse conditions (protect from potential threats) and transformation conditions (use the opportunities emerging from trends, technologies, and cultures you see scaling). There will never be a one-size-fits-all visionary idea, but these ideation sessions can at least nail down what you want to achieve and how it serves your business, customers, or mission.

Think about how much time you've dedicated during the foresight stage and all the activities you've done extrapolating trends, implications, and scenarios into the future—you've already been *living* in the future this whole time just by doing those activities. By now it should be starting to become internalized. Prior to the workshop, you should already be immersed (in your head, at the very least) in what the future could look like. If people are not aware of this, remind them that they've already been living in the future. Now it's time to design something in *that* future.

But relying on everyone's imagination and cognitive capacity to hold these versions of the future in their heads is not enough. As you begin preparing the invitations, documents, and materials for the event, you'll want to make sure that they are also *designed* in a way that they are easily digestible for your audience, allowing the audience to effectively utilize them for ideation. If you have the time, invest it in polishing your artifacts and thinking about when, how, and where they'll be used in the workshop. Do you want to send information ahead of time? Or have a brief meeting to remind people what you'll be doing and answer questions or doubts? (You could use a premortem, as mentioned in Chapter 3.) And consider what would be the most effective way to mentally (or even physically) put people into the future so they will be ideating while immersed in the conditions you've extrapolated. Creating a belief system that they are living in the future can be as simple as showing them illustrated scenario cards, showing them a video, or dressing up the workspace with props, headlines, or other artifacts to illustrate the scene or world they have to contend with.

Next, I'll share a few suggestions for when you are preparing for the design futures ideation workshop:

- Collect and prepare your research artifacts for ideation.
- Identify who will be participating and prepare them for the session.
- Design the ideation experience (selecting the methods).

Dystopian Provocations

Futur2, a small consulting firm and Critical Design studio in Berlin, Germany, integrates Futures Thinking at different points in its process. To prepare workshop participants for ideation, the studio asks them to do homework. Both as a data-gathering activity and to instill a level of investment, participants could be asked to complete a "future vision statement" describing the future of their company or to come up with a speculative solution to a personal problem. This exercise allows participants to share and have a voice, which ultimately transfers to part of the knowledge that's used in the workshop. Futur2 resourcefully uses common frameworks like STEEP but tries to put a darker spin on them to get people to think about unintended consequences, black swans (discussed in Chapter 11), and other hidden threats that could affect businesses or society. The studio likes to call its version of STEEP "Black STEEP" and couples that with a set of Dystopian Trend Cards (Figure 10-1) it created. Each card has its own dystopian provocation of the future. Surprisingly (or perhaps not surprisingly), many of the cards' topics, which were imagined in 2020, have since started to come to fruition, quickly dating them as future provocations. Another device Futur2 uses is the act of role-playing in a business context. Either before or during ideation workshops, it might assign each person a role within different ecosystems or the business. When placed into certain situations, they may have to negotiate their interactions, which can stimulate new ideas but which also illustrates how they perceive themselves in a wider system and the effect that has on their business or operational environment in order to make changes actionable.

Figure 10-1. Dystopian Trend Cards, Futur2

COLLECT AND PREPARE YOUR RESEARCH ARTIFACTS FOR IDEATION

What you bring into the ideation session sets the stage for how your participants will be ideating. Consider which artifacts (trend cards, scenarios, reports, futures wheels) are necessary to get everyone inspired. Even if your participants are the same team that has been with you since the beginning, you'll want to remind them about the future world they are designing within. Unless you are doing back-to-back sessions and jumping from foresight into design, people can forget things easily. If you have the time and the space, determine how you want to include those artifacts in your session. Do you want to send them as a pre-read, or should you post them in the room or on your virtual board as references during the workshop (Figure 10-2)? Either way, you don't want people to forget that this is the future you are designing within.

Figure 10-2. PRIMER 18 conference workshop by Elliott Montgomery, Disruptive Modeling for Higher Education. Along the hallways that led to the workshop space, Montgomery posted headlines from the future to prepare participants for a workshop to imagine new models for higher education in a world based on the extraction and provocation of trends affecting technology, the environment, and the political climate that was emerging at the time.

Think about what artifacts are most important. You also don't want to overwhelm people and distract them from the mission at hand. Having too many inputs into the ideation activity can create extraneous debate and turmoil (sometimes this is a good thing, but depending on the time you have, it could be distracting). A simple 30-minute presentation to open the session could be all you need to ramp people up.

Pre-reads

Suppose some time has passed since the last foresight activity, and you are about to begin the design ideation phase: you may want to send out some information reminding your audience of what was accomplished in the foresight phase. Which scenario or scenarios did you all align on? Which implications were the most important to address? Did you go the extra mile and visualize your scenarios as storyboards or images? Providing access to these results before the workshop can be useful for getting everyone into the right mindset for

designing. However, be careful not to overwhelm people with too many artifacts or too much information without context. You may need only a few key pieces of information as reminders. If you are sending out an email before the workshop, you may want to limit the information in the email to make sure people read it. But always accept the fact that not everyone reads their email! Do you have a repository (such as a Notion page or an internal wiki site) where all this information has been stored? You can always include a link to the larger repository for those who want to dive deeper. But be careful when sending anyone into a repository, because that could be overwhelming for them. Either summarizing the information they need to know or highlighting the information for them might be enough.

In-workshop reminders

You may choose to remind people about the future worlds or foresight outputs *within* the workshop itself. This can be tricky, because you may have a limited amount of time. It may take some time to design, iterate, and align on the idea you want to move forward with, so trying to front-load any preliminary information is always useful. However, if you do decide to ramp up your audience within the workshop, make sure to choose the right amount of information, and decide how you want to remind everyone about future conditions or dependencies so that it is both inspirational and informative. It can take some time for people to consume and process information about a world that doesn't exist yet, so make sure you have ample time in the workshop for digesting information, discussion, and brainstorming. Grappling with the unfamiliar can be a test of how well people understand and how much they believe the scenarios you've proposed. If they can't relate to the future, then they'll have an even harder time ideating, with or without these parameters.

The Autonomous Car Quick Start Guide by Near Future Laboratory

In 2015, Near Future Laboratory conducted a Design Fiction workshop at IxDA's Interaction15 conference in San Francisco, California. The objective was to design a quick start guide for autonomous cars based on the premise that, if and when autonomous cars become more ubiquitous, they may need to include an easy setup guide to introduce owners to the basic operation and safety features of the vehicle. Just as in the small pamphlet that might come with a consumer electronics device, the information had to be highly curated and simplified while still highlighting basic functions and what to do in the event of an emergency. To prepare us for the workshop, Near Future Laboratory decorated the workshop space with news headlines and articles (Figure 10-3). Some of them were screenshots of current online magazines and articles on autonomous cars, current events, or trends, while others were fabricated to look like news headlines, articles, or advertisements from popular media sources in the future (Figures 10-4 and 10-5). These props allowed us to immerse ourselves in the future world and understand the conditions of that world through the lens of what was being reported and how autonomous cars were being used by average consumers and services. At the time of this workshop, California was going through a drought—not a frequent event, but it did happen periodically, which led to many water conservation efforts across cities. Thus, there were even articles about future droughts.

Figure 10-3. IxDA Interaction15 conference workshop by Near Future Laboratory: fictional ads and news articles from the future to construct the conditions of a world in which driverless cars exist

Figure 10-4. IxDA Interaction15 conference workshop by Near Future Laboratory: this image mimics a Guardian *article about how commuters' health and well-being might benefit from the use of autonomous transportation*

Figure 10-5. IxDA Interaction15 conference workshop by Near Future Laboratory: this is a fictionalized home page for Little Caesars (a real pizza delivery store) advertising the use of driverless cars for its delivery services

The breadth of information we consumed to inspire our thinking allowed us to place ourselves into that future from various perspectives and prompted us to discuss what might be needed due to the advent of autonomous vehicles, and what might still be happening or unsolved from today. Ultimately, these provocations allowed us to think about potential use cases and the safety risks that might be necessary to highlight in a quick start guide. Teams were divided into focus areas (Figure 10-6), and we developed the topics, language, policies, and procedures that would eventually be placed into the quick start guide (Figure 10-7). After the workshop, Near Future Laboratory translated our work into a more formal (designed and polished) format (Figure 10-8) and included additional concepts such as brand-sponsored features within the car, which considered how brands might partner on technology or other features. What if Amazon started building autonomous cars? And what if Uber, Amazon, and a detergent company partnered on a cleaning feature for the vehicle (Figure 10-9)? How might brands coexist and contribute to the user experience or technology we might find in the vehicle?

Figure 10-6. IxDA Interaction15 conference workshop by Near Future Laboratory: workshop participants were divided into teams and asked to ideate on the information necessary for the quick start guide within specific categories

Figure 10-7. IxDA Interaction15 conference workshop by Near Future Laboratory: images from ideation to determine the information and illustrations necessary for the quick start guide

Figure 10-8. IxDA Interaction15 conference workshop by Near Future Laboratory: the final publication of Helios: Pilot Quick Start Guide

Figure 10-9. IxDA Interaction15 conference workshop by Near Future Laboratory: a page from Helios: Pilot Quick Start Guide *with a fictional detergent brand company called RE-FRESH that partners with Uber and Amazon for an internal cleaning feature in the vehicle*

IDENTIFY WHO WILL BE PARTICIPATING AND PREPARE THEM FOR THE SESSION

Think about who you want to invite to the workshops. A mix of creative thinkers, strategists, experts, and everyone who has been on your team throughout the foresight stage is a good balance. But overall, you want people who are still excited to contribute and dig into imagining new ideas—people who have an entrepreneurial spirit but are also deeply invested in the outcomes of the project. If you have any newcomers, think about how much they need to know before they come to the workshop. Do you need to schedule a separate meeting with them to update and prepare them for what to expect? Understanding all of this as early as possible can be valuable when trying to prepare and conduct the workshop; you could even figure this out during the diagnostic phase as you try to align all the parties that will participate in each stage and activity. If you miss an opportunity to properly prepare someone or are not prepared for unintended drama or conflict, you could end up wasting a lot of time and energy managing the expectations in the room. This could quickly eat up precious time that you need for ideating, sharing, and alignment. Try to be proactive about explaining

the goals of the session. Set expectations for how the session will be facilitated, how long it will take, and the outputs you want to deliver. Showing pictures of both low- and high-fidelity outputs can ease any tensions around how much detail is necessary to describe an idea. Typically, I like to show rough sketches (Figure 10-10) to get people used to the idea of feeling comfortable with drawing, and to remind them that it's about the idea, not the quality of the drawing or words. Once people realize this isn't an art class and that play and exploration are encouraged, it can start to feel less intimidating.

Figure 10-10. San Francisco Speculative Futures meetup and workshop: a participant holds up low-fidelity sketches for wearables in 2030

Preparing people for creative activities is just as important as preparing people for the entire futures process, as discussed in Chapter 4. I say this because, as you are designing the future, it can also be a divergent and convergent journey within itself, one filled with many ideas, debates, and discussions that eventually you want to convene on as a North Star vision or provocation. Design ideation requires that people push their thinking, step outside of their comfort zones, and imagine possibilities without constraint, though the future world you present will have influences and conditions that can act as guardrails and constraints. How might you get them comfortable enough to operate within those parameters while still being exploratory and visionary? Do you need to show them bad examples and good examples? The best way to prepare them is to provide them with enough information about the exercises, make them feel safe and inclusive to play, and remove criticism and judgment from any idea. Everyone should build upon each other's thoughts and construct the best ideas together.

DESIGN THE IDEATION EXPERIENCE (SELECTING THE METHODS)

Designing the future isn't easy, but given the right environment, activities, facilitation, and inspiration, you can break new ground when trying to come up with an innovative idea. Do not rush this phase! If you can afford the time, plan for it well in advance. Try to think about the sequence of exercises, and make plenty of room for sharing out, discussion, debate, alignment, and iteration. Having just one round of ideation may not be enough to get to a provocative output. Think about how many exercises you want to use to explore different aspects of a problem. Sometimes multiple iterations or a variety of formats can be useful for challenging everyone's assumptions and creativity. If you choose to use only one ideation method, consider making time for multiple iterations to allow teams to debate and refine their ideas. In my experience, one round of ideation can sometimes produce brilliant ideas, but it often doesn't. Providing enough space for discussion and further exploration, pivots, or remixing or combining ideas can lead to more provocative and innovative solutions. You may even want to consider breaking up the ideation across multiple workshops or days. This can give people time and space to digest their thoughts and to find more inspiration between sessions. In the next section, I'll discuss a few method mechanics that can be useful for brainstorming. Determine which ones suit your needs in the time you have, and try to put some thought into how and when you conduct these exercises so that you get the best output from your teams.

Ideation Methods

The methods you choose, how you facilitate them, the time you have, and the collaborative dynamics of your team or client will determine the quality and fidelity of your outputs. Do you want a lot of rough ideas that you may need to analyze and refine more deeply later? Or is your workshop about getting those details refined *with* the team itself? You may have only enough time to generate a volume of good and bad ideas and will need to sort through them and refine them in another workshop or within a smaller team later.

If you have done a thorough job of identifying which methods will be most effective and have budgeted enough time, you should be able to get a lot out of your team. Of course, anything can happen, and hopefully you are also budgeting for potential failures, disruptions, politics, or surprises. A good facilitator will not only design an effective platform for exploration but will also have contingency plans in case the unexpected occurs. In this section I'll just mention a few method types, but keep in mind that there are many other examples out there, and more are being developed every day. The method is merely the vehicle within which your team can use a mechanical and conceptual device to derive an idea. Some methods work better than others, but it will be up to you to determine which is best given the time you have. A few categories of method types are:

- Rapid ideation
- Prompts
- Narrative-based prompts
- Headlines and stories from the future
- Immersive experiences

RAPID IDEATION

If you are short on time or just want to jolt the imagination quickly, you can use a number of *rapid ideation* exercises. Rapid ideation methods focus on quantity over quality and allow you to produce as many ideas as possible within a timebox. *Crazy 8s* is a method that was popularized in Jake Knapp's book *Sprint: How to Solve Big Problems and Test New Ideas in Just Five Days* (Simon & Schuster). The idea is to come up with at least eight ideas in eight minutes (Figure 10-11). But you can modify this to be Crazy 4s or Crazy 10s—it doesn't matter; the point is to get your team to generate as many ideas as possible within an allotted time period. This can be useful for stimulating creativity or even as an icebreaker to prime your team members for the more intensive brainstorm later. From there you can combine, iterate, or use the ideas as thought starters to generate more refined ideas.

Figure 10-11. Crazy 8s

The benefits of using rapid ideation exercises include the following:

- They are fast and easy to explain.
- They require very little setup (you just need some paper and pens).
- They produce a high volume of ideas.
- They can be used as a creativity icebreaker.

There are several cautions to consider as well:

- Ideas could feel unrefined and superficial.

- For the best results, iteration and deeper brainstorming are necessary afterward.

- People might not be used to or comfortable with short time frames.

- The exercises could seem rushed or might be perceived as too fast for the quality that's needed.

PROMPTS

Prompts are ideas (words, images, sounds, stories) that are meant to stimulate thinking and creativity. They provide constraints and a variety of lenses that could be used to inspire ideation. The Thing from the Future (Figure 10-12) by Situation Lab (Stuart Candy and Jeff Watson) is a set of cards that have a variety of prompts representing Arc (time horizons), Terrain (place, domain), Objects, and Mood. When placed together, these four cards are meant to inspire ideas with specific parameters. Since the Thing from the Future was released, many other card prompt sets have emerged, including Near Future Laboratory's Work Kit of Design Fiction (Figure 10-13). However, a prompt doesn't have to be a set of cards or a game; any type of image, sound, story, or proposition can be a way to force a stimulus for creative thinking. The key is to provide a way to ignite the imagination with reasonable (or unreasonable) boundaries so that people have guardrails to work within. Prompts can be used to constrain or unconstrain thinking. Used creatively, they can push people in new directions and inspire new and novel combinations of ideas. Combined with rapid ideation or other storytelling devices, prompts can be a powerful way of exploring futures from totally different perspectives.

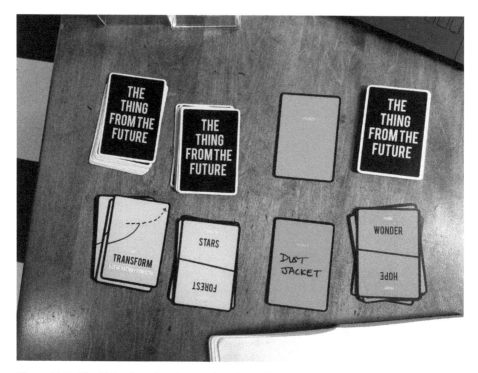

Figure 10-12. The Thing from the Future, Situation Lab

There are several cautions to consider as well:

- Ideas could feel unrefined and superficial.

- For the best results, iteration and deeper brainstorming are necessary afterward.

- People might not be used to or comfortable with short time frames.

- The exercises could seem rushed or might be perceived as too fast for the quality that's needed.

PROMPTS

Prompts are ideas (words, images, sounds, stories) that are meant to stimulate thinking and creativity. They provide constraints and a variety of lenses that could be used to inspire ideation. The Thing from the Future (Figure 10-12) by Situation Lab (Stuart Candy and Jeff Watson) is a set of cards that have a variety of prompts representing Arc (time horizons), Terrain (place, domain), Objects, and Mood. When placed together, these four cards are meant to inspire ideas with specific parameters. Since the Thing from the Future was released, many other card prompt sets have emerged, including Near Future Laboratory's Work Kit of Design Fiction (Figure 10-13). However, a prompt doesn't have to be a set of cards or a game; any type of image, sound, story, or proposition can be a way to force a stimulus for creative thinking. The key is to provide a way to ignite the imagination with reasonable (or unreasonable) boundaries so that people have guardrails to work within. Prompts can be used to constrain or unconstrain thinking. Used creatively, they can push people in new directions and inspire new and novel combinations of ideas. Combined with rapid ideation or other storytelling devices, prompts can be a powerful way of exploring futures from totally different perspectives.

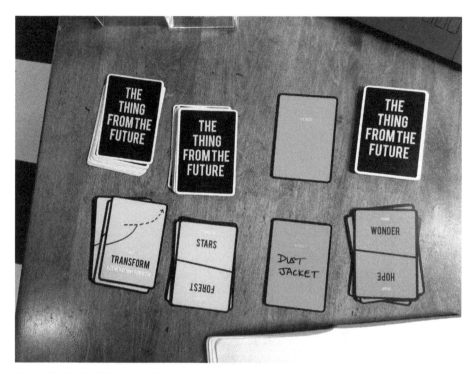

Figure 10-12. The Thing from the Future, Situation Lab

Figure 10-13. The Work Kit of Design Fiction, Near Future Laboratory

There are several benefits to using prompts:

- They're easy to explain and facilitate.
- They can be turned into a game or a competition.
- There are endless combinations of constraints and parameters.
- They can stimulate new perspectives and reframing of problems or solutions.
- You can create your own prompts and define the parameters, categories, or formats.

Some cautions to consider with prompts:

- If you create your own prompts, that could require some time to design.
- They can feel too detached from the actual goal.
- If the prompts are too unconstrained (extreme time horizons) or without certain guardrails, ideas could spiral into absurdity (though that may be what you are aiming for).
- They could seem too constraining.

NARRATIVE-BASED PROMPTS

Prompts can also be used in a narrative format in which some clues are given and participants have to fill in the blanks through storytelling. In the 1950s, a popular children's game called Mad Libs (*https://oreil.ly/K4go_*) was invented by Leonard Stern and Roger Price. It was a booklet of stories with certain words intentionally left out that the players filled in, thus creating wild and sometimes hilarious new stories (Figure 10-14). The same mechanic can be used to create a narrative for futures ideation. This can be used as a prompt to guide people toward the gaps they need to fill in with the ideation process.

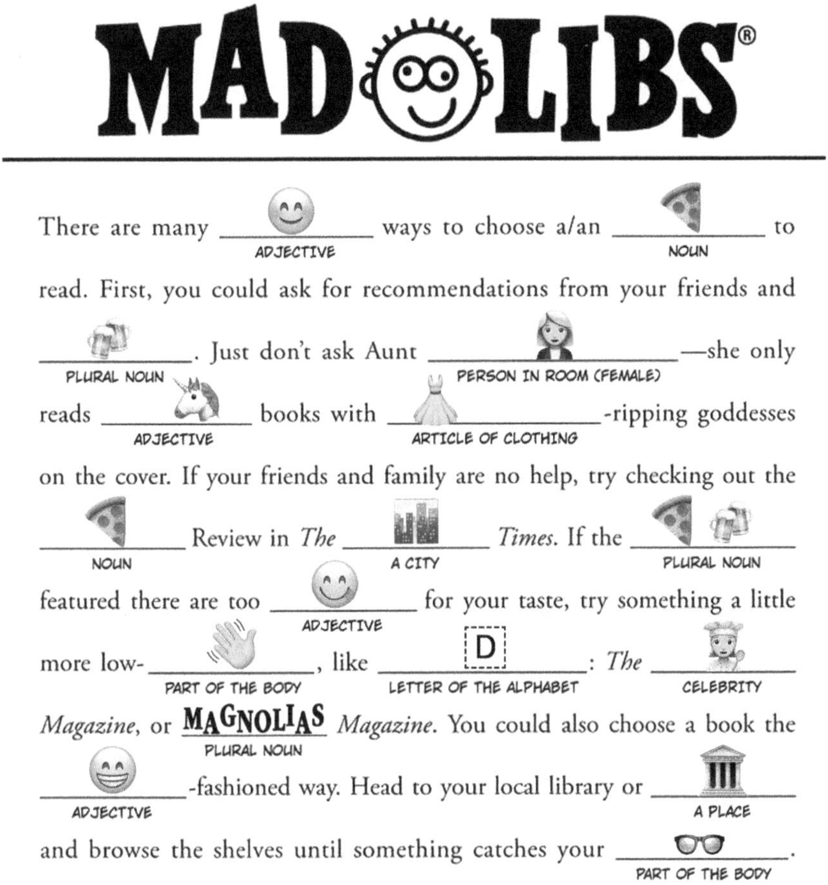

MAD LIBS®

There are many _____ ways to choose a/an _____ to
           ADJECTIVE   NOUN

read. First, you could ask for recommendations from your friends and

_____ . Just don't ask Aunt _____—she only
PLURAL NOUN   PERSON IN ROOM (FEMALE)

reads _____ books with _____-ripping goddesses
      ADJECTIVE   ARTICLE OF CLOTHING

on the cover. If your friends and family are no help, try checking out the

_____ Review in *The* _____ *Times*. If the _____
   NOUN   A CITY   PLURAL NOUN

featured there are too _____ for your taste, try something a little
                   ADJECTIVE

more low-_____, like ___D___ : *The* _____
  PART OF THE BODY   LETTER OF THE ALPHABET   CELEBRITY

Magazine, or **MAGNOLIAS** *Magazine*. You could also choose a book the
          PLURAL NOUN

_____-fashioned way. Head to your local library or _____
ADJECTIVE   A PLACE

and browse the shelves until something catches your _____.
                                          PART OF THE BODY

Figure 10-14. An example of a Mad Libs game

Here's an example of a narrative-based prompt:

It's 2030, and every day I wake up to the sound of (technology) . The first thing I do in the morning is look at my (device) and check my (data) .

Using narrative prompts also gives people a way to ideate with contextual situations. You can detail a day in a future person's life, or you can set up the environment, rules, and conditions to remind your team what it's designing for and why, as in the following example:

After the pandemic, the government passed a law on (policy) . This meant our entire operation had to (change) . This created an attitude of (emotion) . So to compensate and retain our employees we created a (idea) for (goal) which (function) and (function) . This ultimately provided value to (customers) . And we knew this was a successful strategy because we saw (measurement of success) .

Here are some of the benefits of using a narrative-based prompt:

- It provides guided direction on specific characteristics or attributes about a future.
- It uses a narrative format as a storytelling device.
- It can support collaborative thinking (if used as a collaborative activity).

And here are some cautions associated with it:

- It could ignore other important aspects about the future.
- It does not incorporate a visual illustration (unless you ask the audience to create one as part of the task).
- It could be limiting in that it is constrained to preselected stories about the future.

HEADLINES AND STORIES FROM THE FUTURE

Another quasi narrative-based prompt is typically called *Headline from the Future* or *Press Release from the Future* (Figure 10-15). This can be in the form of a news headline, a press release, or another story from the future about your idea. The purpose of the exercise is to help people ideate within the context of a story or from an outsider's perspective *from* the future. When we think about how a news source might report on a particular product, topic, or idea, we are forced to think about how they might communicate its features, its value proposition, its impact on people or the environment, or any other facet that might make it interesting or valuable to read about. These stories also allow you to articulate your idea within a narrative that can incorporate other characteristics of future environments that you may not have thought about.

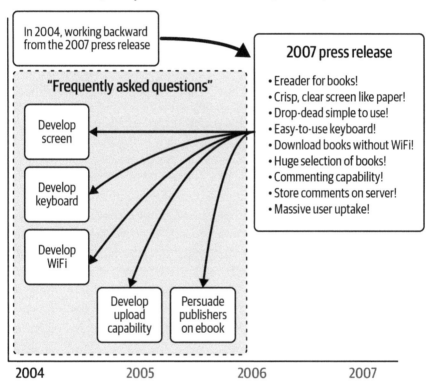

Figure 10-15. Amazon's Press Release from the Future detailing a 2007 press release for the Amazon Kindle (source: adapted from an image by Steve Denning (https://oreil.ly/LCTPK))

The benefits of using a headline, story, or press release from the future include the following:

- It helps to quickly articulate scenarios.
- It provides a context for ideation (news, press release).

- It allows for creative exploration about the narrative of the idea.
- It can include a visual representation either of the idea or of some contextual situation the idea exists in.

The cautions include:

- It could ignore other important aspects about the future.
- It can be limiting in context, breadth, or details.

IMMERSIVE EXPERIENCES

Imagining what the future is like can be difficult cognitively. Holding all of this in your imagination when you close your eyes can take a lot of brainpower. Another way to immerse people in the conditions of future worlds and stimulate creative thinking is to literally and physically build the future world around them through designed environments, artifacts, and situations. Then you can ask: how would you behave in this future? What might be missing that would make this a comfortable scenario? What do we need in this environment or scenario to succeed, or how might it fail?

Many types of immersive and interactive exhibitions are intended to provoke your thinking about the future, but depending on the intention and curation, you aren't always asked to design a response at the end of an exhibit. The same model, however, can be used during the ideation phase in which you can create a physical environment filled with stories and artifacts to determine what's missing or what needs to be designed to fit into a world or respond to an issue or a threat. An experiential future can be an ideation activity within itself (design the experience, the environment, and the artifacts), as well as a way to prototype the future (have participants experience the scene so that they can respond or provide feedback about it). Thus the experience becomes the design problem and the designed artifact itself.

As mentioned in Chapter 8, immersive experiences are not only a great way to depict scenarios but also a great way to visualize, prototype, and experience the future. Stuart Candy and Jake Dunagan, who coined the term *experiential futures*, are some of the earliest pioneers in creating immersive futures design. In a series of exhibits launched in 2010, Dunagan, who was disgusted by the research he was doing at the time on the impact of plastic on our oceans, our environment, and our health, worked together with Candy, artist-producer Sarah Kornfeld, and marine scientist Dr. Wallace "J" Nichols to create *Plastic Century*

(Figure 10-16), an interactive exhibit at the California Academy of Sciences that consisted of four large water coolers filled with plastic debris to represent the total amount of plastic produced and existing on earth at four different points in time.

Figure 10-16. Plastic Century, *Jake Dunagan (photo by Mike Estee)*

Dunagan, who is director of the IFTF's Governance Futures Lab, explained his project in an interview:

> Our goals for the project were threefold. First, we wanted to show, in a compelling way, the exponential growth of plastic production over the last 100 years, and project the levels into the future if no interventions are taken. Second, we wanted to demonstrate that water, pollution, and humans are intimately connected; plastic doesn't go "away." And third, we sought to trigger the "wisdom of repugnance" and install a level of disgust that will stick with people beyond the initial experience.
>
> We are trying to recalibrate people's reality, but there is room for playfulness, ambiguity, and what we might call scientifically rigorous poetic license. Repugnance is a powerful string to pluck, and it should be done

delicately and sparingly. But if done right, it can have a tangible effect on how people see the world and how they act.

Dunagan and Candy's work sought to provide a different objective view on how we experience speculative or futures work by literally wrapping the future around viewers rather than solely providing artifacts or diegetic prototypes. Thus the experience, its setting, artifacts, dialogue, and contextual information in 3D and 2D, become the design. They later proposed the *Experiential Futures Ladder* (Figure 10-17), which builds on the layers of how one experiences future scenarios, from the abstract to the concrete.

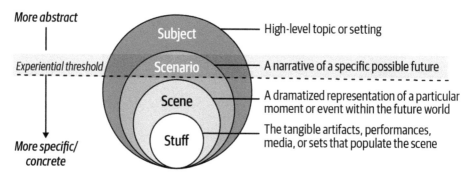

Figure 10-17. The Experiential Futures Ladder (source: adapted from an image by Candy and Dunagan)

As a design approach, this type of experience can be more time consuming and require more thought around everything from the environment, the actors or hosts, and the artifacts and setting to the particular journey or context you want to set for your audience. While an immersive experience can be executed with low-fidelity prototypes or set pieces, the level of fidelity you choose in creating the future world can have a lot of impact on how people believe and experience that world. Whether this is a room, workshop, exhibition, or sequential narrative, designing an immersive experience requires a much more detailed level of design, since there are more factors to consider about how you set up the scenario (what does your audience need to know before or during the experience), what the audience will do (how will they interact with the experience), and what they will learn (how do you measure impact and feedback). While some of these experiences may feel like theater or an art exhibition, they don't necessarily have to feel like something you just view (as you would art). Placing more attention on the interactivity and participatory aspects will allow you to design a living prototype that can enhance the designed product or service you may be seeking to validate or stimulate discourse about.

There are several benefits to using an immersive experience as your ideation method:

- It allows you to understand what the future experience might really feel like.
- It can be a fun, engaging, and interactive experience.
- It creates a space to allow others to provide feedback and test ideas.
- It provokes thinking about additional dimensions, contexts, or situations about a future idea.

There are cautions to consider as well:

- An immersive experience can take more time, funding, and effort to prepare and design.
- It requires more planning.
- If not done properly, it can be seen as theater and may be perceived as extraneous.

ERA: An Immersive Futures Experience

In October 2023, I collaborated with Catalan restaurant Espai Puntal, Berlin-based studio N O R M A L S, Futurity Systems, and other local artists to produce an immersive futures experience in Barcelona titled ERA. Conceived as a parable to reflect on future food systems on Earth, it presented a dire situation where minerals for soil rejuvenation must be sourced from space, and invited guests to assume the role of blue collar workers sent by the Catalan Space Agency to farm a nearby asteroid. Cedric Flazinski and Régis Lemberthe of N O R M A L S, who drove much of the creative strategy, developed activities based on research they did on asteroid farming and space food. Informed by conversations with several NASA advisors, the project crystallized into a two-hour experience during which guests enjoyed a symbolic last meal on Earth before embarking on a make-believe spaceship to a phosphate-laden asteroid. There, they performed space-specific activities (Figure 10-18) and faced a crisis that might bring them to consume their own "harvest."

The event used theatrical techniques to place guests on a speculative future mission that provoked them to think not only about what life might be like as a space worker, but also to reflect on the motives behind space work. Echoing the possible future of humanity as a spacefaring civilization, it aimed to surface analogies between Earth-specific challenges—food scarcity, climate impacts—and the development of new space programs.

N O R M A L S, who are well-known for their particular style and approach fusing fiction, design, and futures, have a diverse portfolio where performance is one of the key components of their work (Figures 10-19 and 10-20). In my interview with them about their process, they said:

Some proposals create a binary mental space that is either "I love this!" or "This repulses me!" The first fails to ask further questions, while the latter sparks instant distrust. For a fiction to be useful is a matter of balance: we must find the problems within the solutions and vice versa. Futures must be explored through the lens of trade-offs: how much might different choices cost—socially, culturally, technically, economically—and are these costs acceptable?

Figure 10-18. Bacteria Preparation Kit, ERA (https://oreil.ly/9pA89). According to NASA, biomining, a form of chemical mining that uses microbial bacteria, is a cost-effective and environmentally friendly way to extract minerals from asteroids. The team proposed the concept of what if space workers could be human bioreactors and actually carry the microbial bacteria in their bodies to the asteroid. Thus one of the activities was for crew members to use the kit to begin gestating the fictional bacteria in their bodies.

Figure 10-19. C D R M X—A Fiction for Mexico City, N O R M A L S (https:// oreil.ly/DfF3K). In 2017, N O R M A L S was invited by Mexico City by LabCDMX to discuss and imagine the future of the city, and of one neighborhood in particular: Xochimilco. Through a lecture and workshop, the team convinced the audience they were part of a fictional relocation agency called CDRMX, an ambitious relocation project combining autonomously 3D-printed housing and universal income (in photo Régis Lemberthe as the fictional economist Rajesh Laghari).

Some proposals create a binary mental space that is either "I love this!" or "This repulses me!" The first fails to ask further questions, while the latter sparks instant distrust. For a fiction to be useful is a matter of balance: we must find the problems within the solutions and vice versa. Futures must be explored through the lens of trade-offs: how much might different choices cost—socially, culturally, technically, economically—and are these costs acceptable?

Figure 10-18. Bacteria Preparation Kit, ERA (https://oreil.ly/9pA89). According to NASA, biomining, a form of chemical mining that uses microbial bacteria, is a cost-effective and environmentally friendly way to extract minerals from asteroids. The team proposed the concept of what if space workers could be human bioreactors and actually carry the microbial bacteria in their bodies to the asteroid. Thus one of the activities was for crew members to use the kit to begin gestating the fictional bacteria in their bodies.

*Figure 10-19. C D R M X—A Fiction for Mexico City, N O R M A L S (https://
oreil.ly/DfF3K). In 2017, N O R M A L S was invited by Mexico City by LabCDMX
to discuss and imagine the future of the city, and of one neighborhood in particular:
Xochimilco. Through a lecture and workshop, the team convinced the audience they
were part of a fictional relocation agency called CDRMX, an ambitious relocation
project combining autonomously 3D-printed housing and universal income (in photo
Régis Lemberthe as the fictional economist Rajesh Laghari).*

Figure 10-20. C D R M X—A Fiction for Mexico City, N O R M A L S (https://oreil.ly/DfF3K). Low-res 3D printed housing—one of the fictional architectural programs. With a desire to preserve their neighborhood history and, just as importantly, their personal stories, the LabCDRMX started a city-wide building scan program, generating an archive of textures and reliefs to be reprinted in the freshly built districts. The process, however, involved resolutions too low to capture all the details, and gave birth to new low-definition architectural aesthetics.

Using Illustrations for Future Visioning

While words can be powerful descriptors, an image can say a lot and may provoke many other thoughts, concerns, or details that can expand on your initial proposal. It's always a good idea to encourage your audience to use illustrations, low-fidelity drawings, infographics, virtual models, or images from the internet, or even to model something physically in 3D using LEGO building bricks, clay, or other materials. The more you can see, touch, and feel, the more experiential you can make the future vision for gathering feedback or gaining alignment. If your audience is not familiar or comfortable with drawing, you can either create an environment in which they can feel comfortable drawing with limited tools and low fidelity or hire a professional to help illustrate for you. Either choice is acceptable, as long as you are able to generate some kind of representation of the idea for everyone to respond to.

Visualizing Futures for Deeper Storytelling

Ellery Studio, a futures and design consultancy in Berlin, Germany, uses a particular methodology for envisioning the future for its clients. Deeply rooted in infographics and data visualization, Ellery uses both hand-drawn and AI generative art tools to help its clients learn about pressing issues and think about what futures they want. A concept Ellery calls the Future Booth has evolved as one of its core tools for extracting insights about what its clients want in the future. The concept is taken from the idea of a traditional photo booth, where you can take a snapshot of yourself in the present. A Future Booth (Figure 10-21) is a similar concept, except you are taking a snapshot of the future. It pushes people to think about what that future looks like. Ellery provides professional illustrators who translate what that future looks like to participants in terms of preferable visions and how they see themselves within that snapshot of the future.

Sebastian Plate, foresight and participation expert at Ellery Studio, explained the concept behind the Future Booth:

You want people to think differently, not just in a strategic or trend-inspired way, but in a way where you look at your visions of the future, not in a sense that they are colonizing the future where only their vision exists, but how they see themselves living in that future. Most of the time it's hard for people to formulate visions. They may have read a report or news article about what will happen in the next ten years, but they don't know what they really wish for within that future, and sometimes formulating that is actually the basis for the Future Booth exercise. We started using it for specific target groups such as children. They knew that they didn't like what was going on right now in the world, but they didn't know how to articulate what they wanted. We saw a similar issue with some of the clients and experts we were working with. They were able to create perspectives, articles, and reports about the future, but they didn't always know what they wanted to see. This exercise helps put it into context, and we can make it very visual. Visualization helps us a lot because it makes things tangible. You can point at it with your finger and say, "This is what I like or want," or "This is what I don't like," rather than having to describe it or rely on what's in peoples' heads.

Figure 10-21. Future Booth, Ellery Studio

In a conversation I had with Bernd Riedel, a cofounder of Ellery Studio, he explained the power of analog sketching to visualize the future and the need for stronger visualizations about the future that are accessible and invite people to participate in the conversation:

Illustrating visions with playful drawings also lowers the bar for accessibility. Too much detail creates a barrier for participants in that they feel like already a lot of work has gone into the vision and it can't or shouldn't be changed. If there are too many details, people can't see where they can have a voice or opinion because they might feel like someone has already done the hard work of thinking about the design or the gaps. But through saying less in an illustration, it invites room for more input, discussion, and transformation. Using playful imagery—cartoons, comic-book-style characters, or even LEGOs or clay—people feel like play is OK. It allows them to interject and contribute with less anxiety or formality, thereby stimulating more creativity and exploration than if something was fully designed in a photorealistic format.

When we attend climate or science communication conferences, experts approach us, usually because we provide creative outputs. When engineers, policymakers, and lawyers think about the future, it is mostly an extension of the present. It comes with the trade. Some say, "I struggle to see any positive images about the future. I want to put my vision out there." This where the potential of climate resilience is the strongest. Using the typical stock photo of the two hands holding soil with a sapling growing out, it does send a message to decision-makers. So we ask the experts to create their vision of the future with us by simply answering a few questions. With generative AI tools, we can now visualize their desires in seconds, and tweak it whichever way they please. Here is an image you can show, and that creates meaning.

Ellery Studio's ability to utilize hand-drawn imagery as well as infographics and other formats of data visualization has been a key to its success as a strategic consultancy for energy and utility companies in Germany and Europe. Ellery also relies heavily on printed images and books as a way to literally put concepts, ideas, and visions in the hands of its customers. While Ellery does create a lot of digital imagery, it has discovered that books create the most joy for its customers and are the most requested items (Figure 10-22).

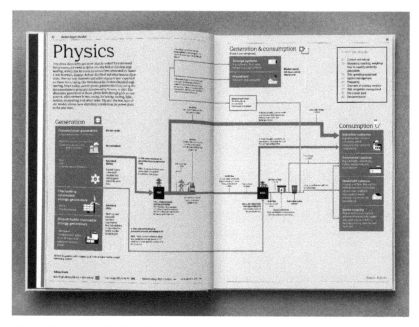

Figure 10-22. Sun, Wind & Wires—Atlas of an Energy System in Transition, *Ellery Studio. Graphic depicting the physical energy transfer between stakeholders within the German energy market.*

Using Generative AI Tools for Future Visioning

Today there are new AI generative art tools such as Midjourney (*https://oreil.ly/ 8pQxP*), DALL-E (*https://oreil.ly/Fic6E*), and Stable Diffusion (*https://oreil.ly/ aU81W*) to help us depict visions of the future. The rise and rapid adoption of these tools, as well as the quickly improving fidelity and realism they provide, has proven that they'll become an asset to artists, designers, and futurists alike. It used to take hours (or a professional illustrator) to illustrate or visualize a scene or speculative object, but now it can be done in minutes with a well-articulated image prompt. In 2022, Magda Mojsiejuk, creative director and lead artifact designer at Futurity Systems, developed a futures magazine called *InTense*, which focuses on selected future topics and includes images generated entirely by Midjourney. *InTense* highlights the vast capabilities of generative AI tools, which can provide you with the ability to generate and play with new concepts that have never been seen before (Figure 10-23). The magazine acts as both a playground

for speculation and a tool and product for Futurity's clients to imagine and distribute ideas within their organizations.

Figure 10-23. NutriOrbs, from the magazine InTense, *Futurity Systems (photo by Magda Mojsiejuk)*

Once you've completed the design futuring stage and have an aligned vision, product, service, or initiative of the future, you'll want to start polishing that vision into a format that can be distributed with your report. If this was the final deliverable for the project, you can stop here and start to collect all your research and artifacts, scenarios, and visions and design the report or presentation. As always, take the time to discuss and understand who the audience will be for this presentation, and design the narrative and information accordingly for maximum impact. If this is the vision you want an entire organization and community to align on, you'll want it to be exciting and inspirational. If you are working on a project where this vision requires a strategic roadmap to build toward, then you can move on to the next section, where we'll discuss how to strategize toward those future visions. This is a critical final stage in making sure your futuring is actionable and measurable and has a plan or project to make it a reality. To sum up all the types of methods and mechanics, you can refer to Table 10-1 as a quick guide to which methods you may want to choose for your situation. Remember, however, that each situation is different, and a method that seems as if it would be ideal and would move quickly might instead end up taking longer than usual. Use your best judgment, and with more practice you'll get better at choosing what works best for each occasion.

Table 10-1. A summary table of method types, use cases, and average time to facilitate

Method type	Use cases	Average time to facilitate[a]
Rapid brainstorming	• Best when you don't have a lot of time • When you need to generate lots of ideas quickly • Silent individual time (allows everyone to participate) • Low preparation	*Fast*—This exercise can run very quickly. However, due to its rapid nature, the ideas it generates could feel superficial, half-baked, or absurd. This isn't bad! But you may want to have more than one iteration to refine your output so that it will more elegantly represent a vision everyone can agree is worth pursuing. I typically give anywhere between *10 to 15 minutes to ideate.*
Games and card prompts	• Group interactivity • Use constraints to push creativity • Abductive thinking • Medium preparation	*Medium*—The time to facilitate this will depend on whether you are using a prefabricated game, set of cards, or prompts or are making your own. Time to facilitate will also depend on how many iterations you want teams to have. Typically I give around *30 to 40 minutes to ideate* around a set of prompts. This can be done as a group or individual exercise.

Method type	Use cases	Average time to facilitate[a]
Narrative storytelling	• Utilizes scenarios and other environmental factors • Considers a day in the life • Need to set the stage for an idea • Low preparation	*Medium*—The time to facilitate this will depend on the amount of information you want a team to write for a narrative or storyboard. If you just have one sheet with a headline/title, description, and image, it can be done in *20 to 30 minutes*.
Immersive experiences	• Systems thinking • Physical role-playing • Moves people around the room • Reenergizes • Lots of preparation	*Long*—This method takes the longest amount of time due to preparation of the artifacts and environment. Depending on the level of fidelity and type of experience you want to create, this can take *several days to prepare*. And depending on the activities or stories you want participants to engage with and how you want them to use the environment to ideate, you could spend 1 to 3+ hours facilitating the actual experience.

[a] Time to facilitate these methods may vary based on your group size and time. Be sure to also allot time to iterate, share out, prioritize, and align on final ideas.

In Closing

For designers and design leaders, the designing futures stage can be a very exciting moment in which we're able to exercise everything we do today with the addition of radical new inputs about future worlds. This stage allows us to imagine and bring complex scenarios and situations to life so that others can touch and experience them. The amount of time you spend in this stage will directly correlate to the quality, clarity, and novelty of your output. Plan ahead, choose your methods wisely, and try to stir up excitement as you align everyone on the future vision. Whether for provocation or for a strategy, design should be treated just as seriously as any other phase of the futures process. With the right approach and impact, it will become the grand vision that an organization can strategize around. In the next chapter, I'll discuss how to plan a roadmap toward that North Star vision and how to make sure everything you've done until now is measurable and actionable—this is part of the final stage of futures work, designing strategy.

Designing Strategy

The final stage of futures work ties the future to the present through the development of a strategic roadmap that can lead you toward your future vision. This section provides some basic methods for developing a long-term roadmap and explains how to set up measurements of success across time, tasks, and deliverables. Futures work is inherently speculative to some degree and can be dismissed as a thought exercise unless you commit a strategic plan to execute on your vision or North Star.

In this stage, we'll delve into the following:

- Chapter 11, "Strategy"
    — Prepare for the strategy meeting or workshop
    — How to use backcasting, a method to kick-start roadmap planning
    — Short-, medium-, and long-term goal planning
    — Measuring success
        — Indicator monitoring
        — Innovation accounting
    — Implementation and delivery of the strategy
- Chapter 12, "Integrating Futures into Your Organization"
    — How to integrate Futures Thinking into an organization from the inside (as an in-house employee) or from the outside (as a consultant)
    — How futures work can fail or succeed

Designing Strategy

The final stage of futures work ties the future to the present through the development of a strategic roadmap that can lead you toward your future vision. This section provides some basic methods for developing a long-term roadmap and explains how to set up measurements of success across time, tasks, and deliverables. Futures work is inherently speculative to some degree and can be dismissed as a thought exercise unless you commit a strategic plan to execute on your vision or North Star.

In this stage, we'll delve into the following:

- Chapter 11, "Strategy"
    — Prepare for the strategy meeting or workshop
    — How to use backcasting, a method to kick-start roadmap planning
    — Short-, medium-, and long-term goal planning
    — Measuring success
        — Indicator monitoring
        — Innovation accounting
    — Implementation and delivery of the strategy
- Chapter 12, "Integrating Futures into Your Organization"
    — How to integrate Futures Thinking into an organization from the inside (as an in-house employee) or from the outside (as a consultant)
    — How futures work can fail or succeed

Strategy

You can think about the future all day, but without some kind of plan for what to do with that visioning, you could be left with merely a "thought exercise." Why do we even think about the future if not to incite change in the way we live or plan today? This is why strategic planning is an essential part of this process, providing an ability to literally create the future we want or avoid the future we don't want. Strategy in the futures process is a way to tie everything you've done into an actionable roadmap. Even if it's just to say, "There's a future we don't want, so let's discuss a contingency plan for how we might respond if it happens." That strategy, whether a multimillion-dollar investment or a list of procedures, is still a workable plan to face the future head-on with confidence. But let's quickly define two words we might use to describe how you plan a strategy—*strategy* and *tactics*. The *Oxford English Dictionary* (*OED*) defines strategy as "a plan of action designed to achieve a long-term or overall aim." It's a plan, path, or roadmap—the tasks, milestones, or steps that must be met or completed to reach a goal at some point in the future. The *OED* defines tactics as "the particular method you use to achieve something." Essentially, the strategy is the long-term plan, and you might employ a number of tactics or methods to achieve that plan. Strategies can change based on the success or failure of your methods. And even tactics have strategies embedded in them to make them work, but essentially every tactic is tied to a greater strategic plan and goal. In this chapter, I'll discuss what to consider when setting up the strategic stage of work, where you can take your North Star or future vision(s) and plan a roadmap consisting of measurable tactics to get there. And within the planning of this roadmap, I'll discuss a particular method called *backcasting*, which is a way to determine what that roadmap looks like and the tactics (milestones, infrastructures, resources, and requirements) necessary to manifest that future vision.

Preparing for Strategy

So how are futures roadmaps different from any other roadmap? They aren't, really. Who is responsible for getting us there? Make sure you choose the right team configuration, one that's truly excited and aligned on the vision. How will we know whether we are succeeding or failing? By setting up agreed-upon metrics. And how might a new strategy conflict or align with a current project? Through careful, inclusive, and detailed planning. The process of planning these efforts isn't really that different from traditional strategic planning. It's part logistics, part collaboration, part financial planning, and part sheer excitement and motivation. One difference you might find is that you will be monitoring an additional set of metrics, trends, and patterns to make sure you are on track.

But planning and nurturing this stage will be an art in itself due to the complexities of how organizations work and the politics, operational parameters, stakeholders, and funding that are necessary to execute on a strategic initiative. In many cases (at least in today's climate), futures professionals are hired to assist teams or organizations in envisioning the future. They facilitate the process, develop scenarios, and generate perspectives, visions, or innovations. After the report or presentation is submitted, however, they aren't always involved in the implementation process. In the best-case scenario, they might witness a shift in the company's strategy. The report might be used in an internal campaign to inspire employees, or it might be released publicly to show off the company's visionary aspirations (much like Apple's Knowledge Navigator video, mentioned in Chapter 6, or AT&T's "You Will" Campaign, discussed in Chapter 1). But at other times, it can become all smoke and mirrors. The company might boast about its pioneering visions but never actually implement the strategy.

As both a consultant and an in-house worker, I've unfortunately seen many of these cases, and it can be a bit disheartening to realize that all your hard work was ignored and never even saw the light of day. Sometimes, that was the intention all along; a future visioning strategic project may have been started just for optics. There are a number of reasons that a strategy might not be implemented, though, including shifts in funding priorities, changes in interest, another competing strategy, some anomalous disruption, changes in leadership, refusal by other business units to adopt the plan, or a disintegration of excitement or interest (I'll dive into this topic in Chapter 12). This is why it's useful to be involved in the strategic planning process and in its implementation as well.

Referring back to the discussion in Chapter 3 of conducting a diagnostic to gather artifacts and an understanding of how your client or team practices strategy today, that information will come in very handy at this stage. By leveraging similar documentation, reports, visualizations, meetings, or other stakeholders who are typically involved in this process, you can model this stage to the way they do things today so that it's not all that foreign to them. Of course, the inputs, content, and time horizon may all be a bit different, but using what's familiar can help get everyone on board and help them be more easily engaged in this process.

As a designer or design leader, you may or may not always be involved in these final discussions. If you are lucky enough to be involved in the planning and implementation process, it can be enlightening; not only will it provide you a voice in the decision making, but you'll also learn a thing or two about the complexities of organizational initiatives. If the client or team does not initially extend you the opportunity to be involved in this stage, try advocating for your participation, or at least try to check in with your client or team every now and then to see how things are going. Offer support, or help even, if it's just a short conversation on the phone. Some companies may really want to implement your strategy but get lost along the way and can't afford to ask for your help, which sometimes leads to falling back on old ideas or ways of working (business as usual) or to scrapping the work altogether.

Planning for the Activities

Much as with any other stage of the futures process, it will serve you well to think about the preparation necessary for the planning session. You've now collected a lot of research, artifacts, and method outputs, and you may even have some speculative artifacts or documentation of immersive experiences you've created that have garnered some important insights about what future you want to create a strategy for. If your team and client have been with you throughout the whole journey and are still excited about the planning phase, then you're in good shape, and you can start this process with the right energy. But some time may have passed since the end of the foresight or design activities, and you may need to remind folks what this final phase is all about. I'll discuss it here as a form of "strategic planning workshop," rather than a standard "meeting," to accentuate the need for activities, negotiation, collaboration, and alignment. If an organization is planning on investing a significant amount of time and money into a futures strategy, surely you'll need to negotiate the planning of resources, budget, and timelines to make sure it happens. Here are a few things you might want to keep in mind:

- Try to use what is already in place to measure success.
- Determine whether new metrics are necessary.
- Invite the appropriate stakeholders.
- Choose the frameworks or methods you'll use for planning.

TRY TO USE WHAT IS ALREADY IN PLACE TO MEASURE SUCCESS

Chances are the organization has already developed a way to implement and measure strategies today. There may be a list of objectives and key results (OKRs) or key performance indicators (KPIs) with very specific logic and calculations.[1] These can be useful in determining how you will plan and measure the success of a futures strategy or any related initiatives developed from it. Try asking how strategic initiatives have been planned and implemented before.[2] Are there some criteria they've been using that determine what is a feasible, viable, and desirable vision or North Star? You should also see whether that model can be directly used in this vision work. Later in this chapter, I'll discuss additional ways to measure return on investment (ROI) in the near and long term.

DETERMINE WHETHER NEW METRICS ARE NECESSARY

In some cases you may need to develop new ways of measuring success, whether with new OKRs, KPIs, or other measurements. You may want to measure the success of the process itself (Were we able to transform their vision of the future? Were people informed, engaged, and excited to walk through the process?), or the success of the aligned vision and creation of an actionable plan (Were executive leaders excited and involved in creating the strategic roadmap?).

Once a long-term roadmap with actionable steps is created, you can determine what needs to be measured to make sure you are meeting the organization's goals within this strategy and all the tactical steps along the way.

---

1 As defined by Rich Sparks in his article "OKRs: The Ultimate Guide to Objectives and Key Results" (*https://oreil.ly/P6jzB*), an OKR "is a popular management strategy that defines objectives and tracks results. It helps create alignment and engagement around measurable goals. Introduced and popularized in the 1970s at Intel, it has since spread throughout technology companies as a way to help employees understand and be engaged in an enterprise's charter."

2 The diagnostic can help answer these questions early. If you've discovered how an organization currently does strategy (roles, formats, meetings, roadmap planning, delegations, and measurements), you can assess those methods now and decide whether they are useful (most likely some of their metrics will be), and you can leverage them (since they're already familiar with them). See Chapter 3 for in-depth discussion of the diagnostic phase.

Measuring the success of the process is just half of the picture, but connecting the process to measurable results fortifies the foresight pipeline. If you need to create a new way to measure success, make sure everyone is aligned on how it is described and calculated and the value it has for the project, teams, or organization.

INVITE THE APPROPRIATE STAKEHOLDERS

Who is typically invited to strategic planning today? Hopefully you've discovered this information in your diagnostic. But maybe it's a different set of people or a new set of stakeholders who will need or want to be involved. Some of this planning might need to involve the organization's or client's resources, executive assistants, or other individuals who are typically in charge of organizing this kind of meeting. Think about who is necessary and who isn't. Consider the stakeholders who can provide the support necessary to fund the effort and have teams and resources that can help with implementation of the strategy. A few to consider are:

- Executive leadership (CEO, CTO, CMO)
- Subject matter experts/domain experts
- Analytics or data science
- Strategists
- Technologists
- Finance
- Engineering
- Product, program, or project management
- Design leadership
- Designers
- Quality assurance
- Marketing

You may even want to have a variety of meetings to discuss the foresight and design outputs before engaging in strategic conversations. Prior to the strategic planning session, you may want to have an initial meeting with executive leadership (C-suite executives) to present your findings, tell the story of how you arrived at this future vision, and describe the details of the goal, product, service, or initiative you want the company to invest a strategy in. Getting upper management informed and excited about the future is critical to getting the top-down support you need to carry on. Getting top-down and bottom-up support should cover all your bases, but rallying all the people who will act as key sponsors and supporters of the strategy could take some time. Try to leverage anyone who has been through the process with you. They will likely be more invested because they've seen the work and have helped generate the ideas that the organization will eventually be investing in. If you're working as an external consultant, it's likely that individuals within the client organization will also have an understanding of the culture, vocabulary, key roles, and dependencies required for advocating the implementation. If you're working from the inside, you may want to rally key leaders and team managers who will be responsible for implementation.

When advising individual contributors—designers, engineers, analysts, and so on—you may want to be strategic in how you inform them about the outcomes of the futures work. If they haven't been involved in the process at all, this information could be surprising or even threatening to their current initiatives. Do your best to cautiously update them on the process and outputs, and try to inspire and excite them about the future vision the team has developed. Exposing as much as possible of the research and synthesis that went into these concepts can show them that there was a process involved (i.e., it wasn't just the C-suite dreaming in a vacuum). If you have time and room for iterative feedback sessions, show the concepts to others to get feedback. While they may be missing some context, fresh eyes on any problem can always be useful.

You may not win everyone's vote of confidence, but putting effort into getting as many people as you can on your side will only be beneficial when it comes to executing on the strategy. Being transparent and inclusive with members of the organization will also create more advocacy and support for the implementation effort and engender less confusion and complaints from people whom you might need to deliver to.

Even if there are people who weren't involved in the process (or the team decided not to expose certain people to the project for confidentiality reasons), advocating for a culture in which everyone can believe this is a preferable future can only help your efforts. Of course, there are instances in which too much transparency can create skepticism, confusion, debate, or antithetical attitudes (there always are). Try to be prepared for the different types of questions and criticisms you might encounter while sustaining some level of trust and excitement even though there is still much uncertainty (VUCA) and potential failure or fears ahead. This is why the futures mindset, at an organizational and cultural level, is very important even before you start the futures work.

CHOOSE THE FRAMEWORKS OR METHODS YOU'LL USE FOR PLANNING

Planning a roadmap into the future shouldn't be overly complicated or require any special frameworks per se, but there could be political, financial, or feasibility discussions that ensue. It's important to have a good grasp of how to facilitate these conversations. You can get creative in how you organize these workshops and *design* them to create a more interactive experience, as opposed to just having a long meeting about how you'll assign resources and who will pay for it. Making it feel personal so that attendees feel as if they have a stake in the game can be very effective. Consider bringing in the important artifacts to remind stakeholders that you're not just planning a roadmap—you're *designing* the future. That said, I'll discuss one of the most common frameworks we use in futures and strategic planning: backcasting.

BACKCASTING

Backcasting is a way of plotting the path toward the future (or futures) you want (or don't want) and working backward in time from that future to the present to identify any key milestones, events, infrastructures, or dependencies that are necessary for that future to be realized. It can be a useful way to identify additional threats, concerns, or opportunities to make sure you can safely plan your roadmap. It connects the long-term goals to the present day so that you can figure out what's necessary NOW to realize that future. Try to remember that you don't have to set off to build the grand vision all at once. Backcasting can be used to set up small prototypes, experiments, and research initiatives that will allow you to slowly inch your way toward your vision. Along the way you can continue to measure your trends, scenarios, implications, and speculations so you can validate that they are still true or determine if you need to pivot, course correct, or revisit your strategy.

In an article in the *European Journal of Futures Research* (*https://oreil.ly/HKyQd*), Simon Elias Bibri related the historical origins of backcasting:

> *The backcasting approach was originally developed in the 1970s as an alternative to traditional energy forecasting and planning and employed as a novel analytical tool for energy planning using normative scenarios. Backcasting studies concerned with energy dealt particularly with the so-called soft energy policy paths, characterized by the development of renewable energy technologies and a low-energy demand society. At the time, such studies emerged as a response to regular energy forecasting, which was mainly based on trend extrapolation and projections of energy consumption, with a focus on large-scale fossil fuel and nuclear technologies. By developing an energy backcasting approach, the focus became analysis and deriving policy goals.*

Originally, backcasting was designed to be a singular path (*https://oreil.ly/t7Wma*), with one singular future to work backward from (Figure 11-1). However, you can also use backcasting as a cone of possible futures you want to consider (Figure 11-2). The idea is to think about how to connect the short- and medium-term goals to the longer-term goals. Again, this shouldn't necessarily feel like a new mental model, because we are always assessing scenarios in our head and trying to figure out how to get from point A to point B and what we might do should certain events occur.

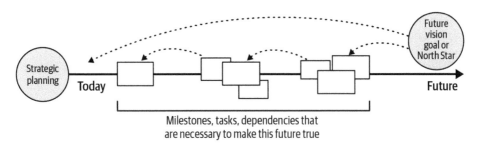

Figure 11-1. *Backcasting*

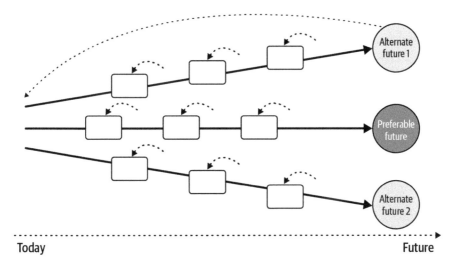

Figure 11-2. *Backcasting multiple futures*

In an interview, Anthony D Paul, Innovation Lab Manager and Design Futurist at GE Transportation, discussed how GE used backcasting in its Innovation Lab to help develop its 10-year roadmap:

We were using our 10-year roadmap [Figure 11-5] as a context-setting tool, to then identify aspirational goals and opportunities as potential desirable futures within the timeline as a future environment. So, for example, a workshop would be to say, if this is the operating environment, what products and services do we introduce within this milestone? Backcasting became an important follow-up exercise because we'd run that exercise many times with different teams—the future opportunity or goal being less important than using it as a cognitive anchor for backcasting—and by backcasting from many of these we'd develop a heatmap of critical dependencies known to support many aspirations. This became a prioritization means for effort that would otherwise continue to be deprioritized today. An easy example might be the need for reliable, network-wide internet. Today, that might not feel like it's important, but for some projects, it is clearly a critical dependency to function and might need to be considered for a particular objective. Thus, these conversations allow decision-makers to see why certain technologies or aspirations are necessary for budgeting purposes.

Going back to our sailboat analogy, we might assess several scenarios or speculations about our uncharted island and work backward to see what route we should take and what we might do if a storm comes or our compass is off. We might even decide to embark on the journey in phases—each day we would travel a short distance, check our data, and continue to make sure it's safe to continue toward our goal. If something were to change, then we would course correct as needed. And if there were any potential threatening unknowns, then we might do more work to try and understand the risk factors of each one. This is a very common process in sailing, as captains constantly have to assess the wind direction and speed and must know exactly what to do to either take advantage of the wind or avoid it. They do this with the help of an arsenal of instruments on their boat and using information that is broadcast from regional weather or marine services. Our instruments are the same as we slowly creep into the future; given everything that is being broadcast by the environment we're heading into, we can better navigate the waters as change happens.

Backcasting as a singular roadmap works just fine, and you can certainly plan out the necessary steps toward one agreed-upon future vision, but it also allows you to create an additional opportunity and risk assessment as you try to detail all the steps of the journey along the way. At every step into the future, you can use a Futures Wheel to assess and troubleshoot areas within the greater context of the strategic plan: what if this happens, or what if this doesn't happen?

You can also map a Futures Cone (see Chapter 4) onto the backcasting map if you want. This turns the backcasting into a form of scenario planning in itself. Each possible future at the end could represent one of the scenarios you created in the foresight stages (unpreferable collapse scenarios, transformation scenarios, or the preferred scenario). In Figure 11-3 I've plotted three types of future visions, each referenced by the original North Star that we want to go toward. There might be a nonpreferable future where that vision fails completely, and another version of the vision where it isn't quite what we expected—maybe it's transformed a few degrees differently, but it's still preferable. For example, if we were planning to start an initiative to address climate change action to reduce our carbon footprint by 80% in the next decade, that might be the ultimate goal. But a nonpreferable vision might be that we don't achieve that goal and reduce our carbon footprint by only 10%, perhaps because the company moves slowly in implementing the right processes, or because it doesn't obtain the necessary funding, or because some other disruptions occur that throw off our plan. This is similar to a collapse or negative scenario. An alternate future vision might be that we get halfway there and reduce our carbon footprint by 40% over the next decade. That's not too bad—some progress was made, though it's not quite the aspirational goal we had set out for in the beginning. Each vision is still within the realm of the goal (reducing our carbon footprint) but slightly different (much like scenarios). Now when doing the backcasting, we can start to map out what happens if we fall off our primary roadmap (achieve 80% reduction) and what we might want to do to get us back on track. If we start to deviate from that plan in the 40% reduction direction, we could choose to just stay that course, or we could decide to pivot again to try to reach 80% in the next decade. These variations of North Star visions provide reference points so that when you are planning, you can work out some of the potential challenges and what to do should they occur. You can add more than three alternate futures—as many as you feel necessary.

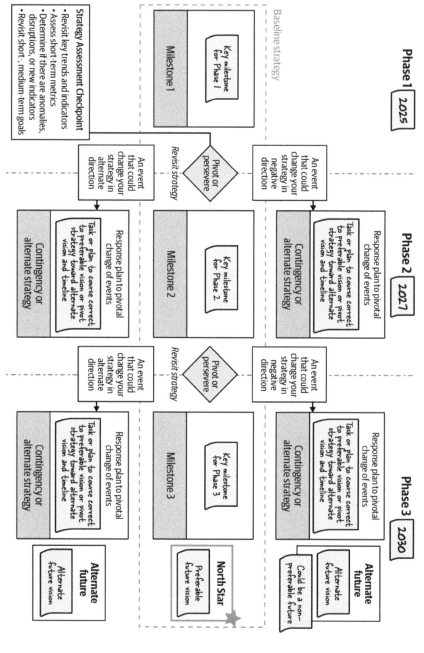

Figure 11-3. Backcasting roadmap considering three possible futures

Some have compared this process to what appears in films such as *Everything Everywhere All at Once, Doctor Strange in the Multiverse of Madness,* and *Avengers: Infinity War.* These sci-fi movies draw on the concept of a *multiverse,* in which decisions or events can create alternate timelines and change the course of histories. In *Avengers: Infinity War* (spoilers ahead), there is a scene in which Dr. Strange is able to look forward in time at 14,000,605 possible futures (Figure 11-4) in the hope of understanding in which timelines the Avengers will win their battle with the antagonist, Thanos. In the end, there was only one timeline that led to a success, but it required a very specific event to occur (the death of Iron Man). This one important event was what enabled the heroes to turn the tide in their favor. Similarly, there might be critical events in your timeline that could be pivotal in changing the tactics, roadmap, and strategy you have planned. Surely you can think of this process this way, and maybe even reference it in your workshop (if it fits the situation and audience). I also use this example when explaining scenario planning. The point is that strategic planning (and Futures Thinking in general) is like having an ability to look forward in time at multiple *possible* scenarios (or multiversal timelines) so that you can plan your way forward. Coming up with your own analogy from pop culture or everyday decision-making situations could be a powerful tool for getting through some of these more complex stages of dealing with the weight of alternative futures.

Figure 11-4. In Avengers: Infinity War, *Doctor Strange looks at 14,000,605 possible futures to find how the Avengers could win the battle against Thanos*

The Future of Freight at General Electric

Around 2011, General Electric began investing heavily in the industrial internet (the Fourth Industrial Revolution), harnessing big data and analytics to improve the performance, efficiency, and predictability of its industrial businesses, such as oil and gas, aviation, and transportation (the business unit responsible for building train locomotives and freight systems). What began as a digital research arm for its Global Research Center of Excellence eventually transformed into GE Digital, located in San Ramon, California. GE Digital was constructed as a dedicated business unit and was meant to pave the way for GE's new digital platforms. Designers and engineers were hired to form an internal agency to build out GE's Predix software platform to consume and report data from its industrial machines and deliver insights to operators and other stakeholders. Around 2019, Anthony D Paul and Ryan Leveille, who started off as interaction designers within the transportation business unit, stood up an innovation lab that would be responsible for helping define the 10-year strategy, utilizing Futures Thinking methods, as well as VR/AR and other emerging tech, to determine the future of GE's freight industry (Figure 11-5). As designers and facilitators who were on the ground working directly with customers and operators, they had a particularly valuable body of knowledge that could be useful in strategic planning conversations.

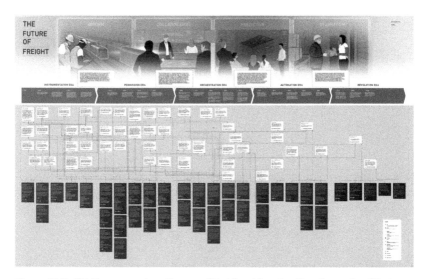

Figure 11-5. GE Transportation, Future of Freight vision timeline (https://oreil.ly/ EPgNn): the 10-year strategic timeline outlines initiatives over five segments (called eras), with each era building on the previous one by investigating, experimenting, and implementing key infrastructure, technologies, and tasks

In an interview, Anthony D Paul explained the origin of the Innovation Lab, its challenges, and its successes:

> *Senior leadership started pulling us in to run workshops for them around how to set up our strategic pillars, because we had a level of insight that they didn't necessarily have in their day-to-day role. We understood all of these pieces and relationships between divisions and between P and Ls [profit and loss] and in different layers of hierarchy within our customer space and tensions between customers and partners. And so we could come in and give a lot of context and add to some of the decisions that were being made—which products do we fund, and which are cash cows. However, products eventually go obsolete. The leadership was also afraid of workforce retirement cliffs where railroading, freight, and mining and a lot of these big old industries tend to be multigenerational employees, so you would have a grandfather and a dad and a son. That's a very different world today to see a lot of more educated generations are coming in and they're coming in from having never been in that industry, and then kind of questioning why things are done the way they are. And that's a big fear for a lot of these customers and industries.*
>
> *So we split off and formed the Innovation Lab to really scratch that itch of how do we have this spectrum of work that seemed to be the polar opposite of what the UX team was doing—using field research to advance incremental priorities to solve the little needs that were aligned with OKRs* **and** *the need to solve the organization's need to solidify a strategic direction for the business.*

In strategic meetings, it's important to get everyone aligned on the vision and looking in the same direction. This can be difficult depending on who is in the room and the competing agendas of teams, stakeholders, or partners. There could be hundreds of projects in play, and when you're trying to align on a central vision that binds the company's strategy together, it can be difficult. There are too many dependencies.

Everybody has their individual itch that they're trying to scratch, but they don't have any sense of whether it does or doesn't align in the same direction. So I began by asking, "What's the performance measurement that you're trying to fix?" and they might say, "Oh, well, we want to reduce dwell time because we have an asset that sits for too many hours a day."[3] "OK, let's snap our fingers, pretend we fixed that and there is no dwell. Now, what's your next performance measure?" "Oh, well, if there wasn't any dwell, then we have this other operational challenge," and that would be the next dependency and the next dependency, and you kind of go through this cyclical loop until you get to an existential crisis point, where people in their mind will suddenly get to a point where, if I take away the near-term constraints that I'm really dwelling on and focusing on today, this is the core vision of what our firm or organization is trying to accomplish. And once I understand that, now I can set a vision for where we want to go as an organization, as a team, or as an industry.

---

3 *Dwell* refers to the amount of time a large asset (such as a locomotive) sits without operating. Dwelling assets can deteriorate quickly, reducing their performance and the longevity of their parts.

And so with each of these teams I needed to get to a point where we were all looking in the same direction. And the quicker I got to that, the quicker I can come back to what was the problem we were trying to fix. OK, now let's align it to that vision: what are the dependencies steps that we need to walk through to get toward that direction? What's the next sprint plan for your software application, what are the critical dependencies that you need to put in because your customers are requiring you to put them in whether they align or don't align, and then which are things that we need to figure out how to begin even though the customer is not asking for it, because it's a critical dependency for us to advance to the next step.

By understanding and looking at that same vision, we can have that level of discussion, and before that we couldn't have that level of discussion. It was just the squeaky wheel problem and whoever was screaming loud enough. So what that ended up amounting to is, as Ryan and I, you know, we kept going through this loop, every time we talked to a new group trying to level them up quickly. And that's where the future for a timeline came in because we needed to have a means of quickly giving everyone all of the context that was in our brains in as little as 5 or 10 minutes to then immediately get to the point where that workshop can be productive and we can advance. And so the future freight timeline was introduced.

Long-Term, Medium-Term, and Short-Term Planning

One of the keys to making futures work is to connect what we want in the future with what we can do today. That means connecting initiatives, projects, dependencies, and results in the near term to the longer-term vision (Figure 11-6). Combining these goals and milestones will allow your audience and stakeholders to realize that the future doesn't need to operate in a far-off time and place.

If an organization is more concerned with near-term results, then focusing on the near-term successes with medium- and long-term goals in mind will keep them feeling more secure about setting long-term goals further in the future. Backcasting will help define these different goalposts for you to reach. But again, the right futures mindset, accepting that change is inevitable and that futures is an ongoing process that requires reassessments every so often, will also allow everyone to be more confident in pursuing the path forward.

Create strategy using backcasting (or any roadmap planning method to identify, short-, medium-, and long- term goals to achieve North Star vision)

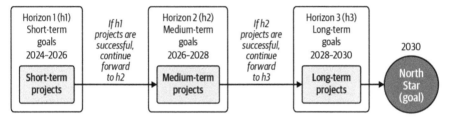

Projects are pointed at the North Star

Figure 11-6. Connecting future vision (long-term goals) with medium- and short-term goals

Set up your medium-term and short-term goals with proper checkpoints and a set of tasks to reassess, remodel, or redo (if necessary) the futures work. Short- and medium-term goals should be plausible, realistic, and attainable and have some set of metrics attached. Your organization or client might even have some very specific goals or metrics that they will need to approve or achieve to make sure the strategy is successful. Use those and include some additional metrics (discussed in the next section) that speak to the transformational impact of the process. For example, let's say your future vision is to develop an AI-driven network that could predict which of your customers want to travel throughout the year and when, where, and how they want to travel. This obviously would require gathering and analyzing a lot of your customers' data and using AI to provide predictions about their behaviors. But if you've never worked with AI technologies before or you don't know where to start to gather this data, you may want to set up stages and goals to reach that ultimate dream.

Consider the story of the Apollo program from Chapter 7, where the vision was to place a human on the moon. However, to reach the moon, NASA had to create several programs to slowly get there step by step: the Mercury program (to determine whether humans could survive in space), the Gemini program (to test maneuvering in space), and the Apollo program, the 11th mission of which landed the first humans on the moon on July 20, 1969 (Figure 11-7).

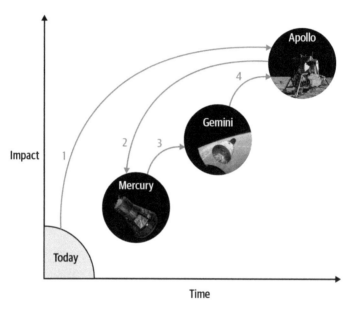

Figure 11-7. The Mercury, Gemini, and Apollo programs that led to landing the first humans on the moon

Consider these examples of a short-term goal:

- Understand the types of personal customer data you need and can access.
- Decide what current policies exist around data sharing or what opt-in policies might have to be created to capture the data.
- Build the tech stack and database infrastructure to store, analyze, and process a large amount of data.

Here are some example medium-term goals:

- Refine the AI algorithms to make sure the application is providing the quality and accuracy of data you need.

- Run a series of tests and prototypes to make sure the prediction algorithms are at a bar of acceptable accuracy.

- Prototype the application with small communities of customers to measure the value of the application and determine if there are features that customers want and need.

And here are potential long-term goals:

- If all goes well and various tests in different regions succeed, scale the application's features.

- Determine additional business models to gain partners, collaborators, and customers.

- Incorporate new technologies or partner applications that could enhance your offerings.

These lists of goals are not exhaustive, of course, but they set the bar for discussion around how you might want to segment tasks and successes (or failures) along the way to your vision. As time marches on, these goals might change if you are successful in the short/medium term, or if disruptions or new technologies are invented, or if you simply fail to reach the market and evolution you hoped for. At least you have a plan and hopefully have thought far enough into the future that you have contingencies for any surprises.

MCKINSEY'S THREE HORIZONS

In Chapter 7 I mentioned Bill Sharpe's Three Horizons framework. There is another three horizons framework that has been popularized by McKinsey & Company as a way to plan out long-term strategies in three segments of time. As a framework, this is also an easy way to discuss and delegate the activities necessary to get from today to your future vision:[4]

> The three horizons framework—featured in **The Alchemy of Growth**[5]— provides a structure for companies to assess potential opportunities for growth without neglecting performance in the present [Figure 11-8].

---

4 McKinsey & Company, "Enduring Ideas: The Three Horizons of Growth" (*https://oreil.ly/aHBZw*), *McKinsey Quarterly*, December 1, 2009.

5 Mehrdad Baghai, Stephen Coley, and David White, *The Alchemy of Growth* (New York: Perseus Publishing, 1999).

Horizon one represents those core businesses most readily identified with the company name and those that provide the greatest profits and cash flow. Here the focus is on improving performance to maximize the remaining value. Horizon two encompasses emerging opportunities, including rising entrepreneurial ventures likely to generate substantial profits in the future but that could require considerable investment. Horizon three contains ideas for profitable growth down the road—for instance, small ventures such as research projects, pilot programs, or minority stakes in new businesses.

Time, as noted on the x-axis, should not be interpreted as a prompt for when to pay attention—now, later, or much later. Companies must manage businesses along all three horizons concurrently. Rather, it suggests the cycle by which businesses and ventures move, over time, from horizon two to horizon one, or from horizon three to horizon two. The y-axis represents the growth in value that companies may achieve by attending to all three horizons simultaneously.

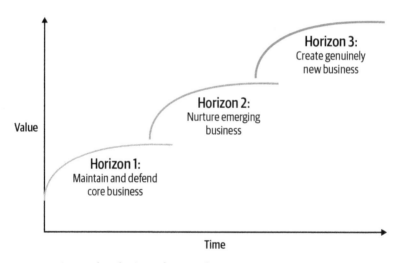

Figure 11-8. McKinsey's three horizons framework

While McKinsey's framework sets forth clear initiatives in each horizon, this structure can be simplified even more by simply adjusting and assigning which activities and tactics are necessary to experiment and validate in Horizon 1 and move the organization's strategy toward what you have defined as the ultimate vision and goal of Horizon 3. If the projects and tasks you complete in Horizon 1 are measured as successful, then you can continue into Horizon 2 as planned; if there are projects that have failed, then you may want to revisit what you hoped to accomplish in Horizons 2 and 3 and adjust as necessary. Fundamentally, this framework is just a way of dividing time into three segments so that you can set markers, measurements, and goals for each period of time as you move toward your vision.

JUMP-STARTING INNOVATION

Once you're engaged in strategic planning, you may discover that the long-term vision you projected 10 years from now can actually be accomplished in less time. By having a deeper discussion about the vision and what's necessary to build it, you expose the finer details about how you will get there. This can be really exciting when you realize you don't need to have as long of a timeline as you originally projected and that innovations can happen faster and sooner. This doesn't mean that the projection work you did was incorrect per se. The process has allowed you to gather the information to understand future landscapes and formulate an idea of where you want to go. You have the authority to "kick the tires" to assess how feasible it is and make adjustments in your plan if necessary. That's how this process leads to innovation opportunities today—innovation that is possible in the far future can inspire innovation *now* merely by having a goal on the horizon as a guiding light.

Unfortunately, it's not always easy to set a vision and rally an entire team or organization to execute toward it. There's always a little fine-tuning of the vision, feasibility assessments, measurements, resource and financial allocations, and teams that need to be assigned to implement the vision. Finding the right balance between getting everyone invested in a future worth building and connecting that to what we can and want to do today is essential in making Futures Thinking work as a strategic approach.

Sometimes your vision is too lofty or fantastic—it might be inspiring and aspirational, but stakeholders can't totally buy into it because they literally can't see the building blocks necessary to make it happen. Sometimes you have to paint each step clearly to show why the vision matters, why it's attainable, and why it's critical to the success of your organization or products. This process of making a future vision grounded in reality can be a sensitive and tedious act of negotiation, discussion, and transformation and can even take several months before you gain enough buy-in to implement. Buckle down for that journey and try to set up the parameters for success as early as possible. Socialize the process with key stakeholders and investors (people high on the Power-Interest

Matrix) so that you can leverage them if necessary during the planning and implementation process.

In an interview with Tino Klaehne, Director of Strategic Innovation and Intelligence at Lufthansa, he discussed the challenges of stakeholders needing short-term horizons they can measure, validate, and execute quickly:

> *The problem of working in a corporation is that you cannot really look far into the future. They want something where they can start now rather than jumping too far in the future. Futures takes a lot of investment in time. We do the work, we prepare scenarios and usually we have to illustrate them and set them up nicely, which takes a lot of time and resources. Organizations build on ideas rather than problems; they want an idea because usually ideas bring in more work.*

Measuring Success

This is one of the most contentious topics I've encountered when it comes to implementing futures work. How can we measure return on investment (ROI) if the future may take years to arrive? Or how do we know if we chose the right future to invest in? How do we know if our strategy or process worked or didn't work? These questions are all great and really stem from our current ways of thinking and methods of measurement, which are primarily based on near-term business goals, financial impact, or customer experience metrics. You *can* measure the success of futures work, and in ways that are similar to how you are measuring your KPIs or OKRs or economic returns today. Tying long-term goals to near-term goals and then measuring progress, outcomes, and changes against them will be key to connecting and acting on a successful futures strategy. And if your organization is interested only in the near-term goals, that's fine, as long as they can also accept some guidance toward a future goal. But using this process to at least get people thinking about long-term implications or opportunities (whether or not they invest in dedicated initiatives toward this vision) is still a success. Opening up a conversation using futures work could be an awakening for people to start thinking beyond near-term ROIs.

INNOVATION ACCOUNTING

Measuring innovation or the impact of innovation processes can be tricky, but it isn't impossible. For instance, startups in their early stages are scrambling to seek validation in the market and a profitable business model. Some startups may not even be deemed "successful" until they have gone public. This isn't

to say they aren't practicing some form of long-term strategic thinking. They obviously have a vision they are working toward through small market-validated prototypes. But when you ask them what their 3-year, 5-year, or 10-year strategy might look like, it could be very blurry still, or dependent on what they can actually accomplish in the first year of existence. However, as I've continued to advocate for in this book, futures can be used anywhere and within any time horizon. If you are a startup and just want to look at the market you're in and try to understand the scope and impact of the trends that could derail or enable your product, you may want to give futures a chance. But startups these days seem to operate around a philosophy that was founded back in the early 2000s and is based on rapid design and development: launch quickly, fail quickly, learn quickly, and iterate based on customer feedback. In his book *The Lean Startup* (Crown), Eric Ries coined the term *innovation accounting* to refer to "a way of evaluating progress when all the metrics typically used in an established company (revenue, customers, ROI, market share) are effectively zero." While Ries's book initially was an investigation into startups and teams, Dan Toma and Esther Gons look at corporate startups and how innovation operates in corporations. In their book *Innovation Accounting* (BIS Publishers) (see Figure 11-9), they defined innovation accounting as

> the process of defining and measuring innovation within an organization. When we are still creating and testing ideas, financial indicators are simply not enough.[6] They offer insufficient data as they don't reflect the progress of the learning process when testing and validating ideas. That is why every modern organization needs Innovation Accounting next to traditional financial accounting. This new system must be in sync with the financial accounting system while mitigating its shortcomings.

---

6 In finance and stock market analysis, there are lagging and leading indicators that analysts watch for trends. The financial media website Investopedia offers this explanation (*https://oreil.ly/2lsld*): "Leading indicators look ahead and attempt to predict future outcomes, whereas lagging indicators look at the past. Some people fixate on leading indicators, arguing that what happened in the past is useless. However, that's not true.... Classic examples of leading indicators include yield curves, new housing starts, and the PMI. Each provide a gauge of where insiders and so-called experts think the economy is heading.... Lagging indicators can only be known after the event, but that doesn't make them useless. They can clarify and confirm a pattern that is occurring over time. The unemployment rate is one of the most reliable lagging indicators. If the unemployment rate rose last month and the month before, it indicates that the overall economy has been doing poorly and may well continue to do poorly. Lagging indicators are very useful at confirming trends and changes in trends. And they are set in stone, unlike leading indicators, which may not always be accurate and can be misleading."

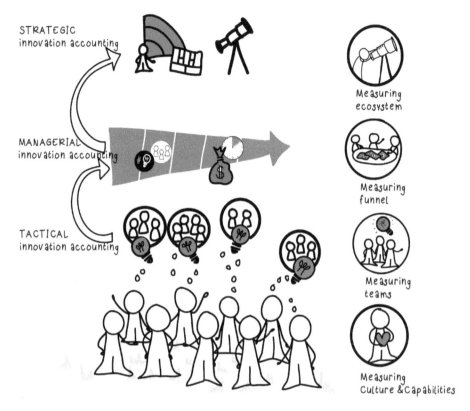

Figure 11-9. Innovation Accounting, Dan Toma and Esther Gons (https://oreil.ly/9Z5mY)

By defining metrics at a tactical, managerial, and strategic level, you can identify indicators that address the levels of change that permeate an organization as innovations are conceived, tested, and implemented over time:

Strategic innovation accounting
> Connected to measuring the product ecosystem. The focus here is on helping a company's leadership examine the overall performance of their investments in innovation in the context of the larger business, connecting to overall strategy as well as decisions on capability and culture.

Managerial innovation accounting
> Connected to measuring the product funnel. The focus here is on helping the company make informed investment decisions based on evidence and the current innovation stages teams are in.

Tactical innovation accounting

Connected to measuring teams. The focus here is on product teams, the experiments they are running, the learnings they are having, and the progress they are making from ideation to scale.

At every level there is a series of metrics that could be applied to measuring changes and successes along the way to sustain an interdependent ecosystem that nurtures the innovation funnel from discovery to sustaining a product in the market. But innovation never evolves without trial and error, and without a system and framework in place to measure ideas, experiments, and impact, you may find yourself throwing too many things over the fence and potentially disconnecting these experiments from the greater vision that you may want to accomplish through the innovation funnel.

Some organizations may demand to see quantifiable impact. Others will see the value of the more transformational, innovative, or systematic effects that futures can provide. But you don't necessarily want to put that thermometer in the distant future to measure because you can't use a time machine to go there to see if it worked. If you want to use traditional methods of impact, you can begin by *prototyping* the first iterations of your future vision today. If this doesn't sound new, that's because it isn't.

When teaching organizations about Agile methodologies—a set of flexible approaches to project management that involve breaking a project into phases and emphasize continuous collaboration and improvement—we often talk about creating the *MVP*, or minimum viable product (Figure 11-10). The MVP is meant to test small hypotheses through low-cost experiments of a bigger idea with the bare-bones minimum number of features it takes to prove whether people want and need it. We sometimes explain it as wanting to build a fancy car but not knowing exactly how to build one. But the goal of the car is to transport you from A to B. That's the vision. An MVP is a project that is lean, costs little to prove (a proof of concept per se), and should be able to achieve the job of transporting you from A to B—a lightweight experiment that still performs the job you want. So you might build a skateboard at first. Once you succeed at that, you might build a more advanced skateboard, or a scooter. Then you might move on to building a bicycle or a motorcycle once you understand more of the mechanics and technologies involved. Each of these experiments, or MVPs, succeed in manifesting the vision of transportation in some way, until eventually you get to the ultimate vision. It may take years to get there. And the vision of the fancy car may have changed based on what you learned over time with your experiments. But what matters is that you had a vision that initiated the experimentation. There is, of course, a scenario in which you abandon the vision altogether. Maybe you got as far as building a bicycle and realized that what you want is not the car but a fancier bike. This is all very similar to the futures process. You have a future vision (the fancy car), but you can test it as a hypothesis with small MVPs along the way. And with every experiment, MVP, or prototype you build, you can measure its success. And even if the technology doesn't exist yet, there could be low-fidelity ways to test "What if?" scenarios to see how it might work if it did exist. These learnings can be used to begin developing the necessary technologies or processes now to lay the foundation for the vision early on.

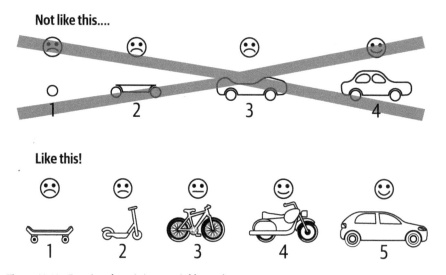

Figure 11-10. Creating the minimum viable product (MVP)

Just because futures work utilizes different tools to develop a long-term vision doesn't mean it's a totally different type of work or goal. There are several typical near-term metrics that could be applied to your early MVPs or prototypes:

Increase in sales, customers, or positive customer feedback
> This is a standard KPI for product success. Your first steps toward your vision might be a small prototype that customers really like. Positive customer feedback or increases in sales or engagement mean you're at least going in the right direction.

Increase in media attention

Whether you are publicly showcasing your future vision or gaining some success through your near-term prototyping or product releases, an increase in attention internally or externally means people are seeing the success and want to know more. This can be inspirational for team morale and can validate that you're doing something well.

Operational savings and efficiencies

As a strategic opportunity, futures has the ability to identify incoming changes in the market that could affect your total spend on various operations (finance, marketing, HR, engineering). With the right strategic planning, you can figure out ways to not only reach your goal but also do so while spending less time, money, and effort to get there. While an operation considers many people, departments, funding, and tools, having an aligned vision that can help inform how an organization runs can allow the organization to plan for potential disruptions, attrition, or shifts in processes ahead of time. (For example, if certain AI tools are set to replace many human workforces, you could assess how much savings in operations you could gain over the years by reducing workforces or retraining them to manage and oversee AI.)

Emergence of or increase in competition or investments

One might think that inspiring competition is bad, but you can also look at it as becoming a market driver. As more people tap into an emerging technology, more is learned. There is more failure and more success to learn from. Competition can also create healthy pressure to innovate and accelerate development. If you're in it to win it, then you'll need to keep an eye on the market, learn from others' mistakes, and continue to look ahead of them so that you can succeed. (For example, if you pioneered an app specifically for children to learn new languages and were first to market, the emergence of copycat apps or models in the market would show that you are a disruptor and an inspiration to others. You could monitor their successes and failures and use those as inspiration to enhance your own application or further disrupt their ecosystems and customers.)

Growth in staff (needed to execute)

If your first stages of strategy implementation are successful, you may start to need more staff to execute. Surely any success will mean the company wants to do more to stay on that path. This human resources KPI can equate to success in the overall strategy as well. More customers and increased sales mean more product is needed, which means more people are needed.

These are all ways we measure the products we deliver today, and they can certainly be documented and affect our bottom line. When tied to the longer vision roadmap, you can claim these successes were due to having a navigational direction and plan toward that vision. Each step and every success you can claim toward the roadmap should be highlighted and added to the list of values you are creating by doing futures work. That's the fastest and easiest way to prove your success. However, there are additional ways to set up metrics other than immediate ROI:

Development of new intellectual property
Futures as an innovation strategy can certainly allow you to identify new opportunities that may potentially lead to new intellectual property (IP). Every new idea becomes something you can capitalize on or monetize (if that's your goal). Of course, it will require some prototyping and testing, but that is what ties today to tomorrow.

Excitement about and awareness of multiple possible futures
Organizations that are stuck in the trenches of current roadmaps could be losing steam thinking about the same problems and the same future day after day. For example, a company that has been tasked with creating the next generation of gaming to increase the fidelity and realism of characters in its games is an ongoing technology and processing power challenge. Yet companies have been working on this for ages, with incremental improvements. But due to mandates to accomplish this goal, there may be little time or resources to focus on anything else that is innovative or forward-thinking outside of their current missions. This can, in turn, lead to attrition, boredom, and a lack of enthusiasm to tackle some of the challenges others in the industry are racing to pioneer. Futures not only will unlock possibilities but also will hopefully pull people out of a fatalistic attitude that there is only one path forward, no matter how exciting or grueling it might be. Analyzing sentiments on social media or within the organization can help provide a nice set of qualitative feedback if your efforts are truly activating people's ambitions about the future.

More holistically driven visions and strategy

With its various lenses and perspectives, futures allows organizations to look beyond their near-term goals and consider impact and influences from other areas (using STEEP for horizon scanning, for example). If your futures work has opened up the organization to thinking about factors other than technology or just its current user base engagement, such as the environment, policy, culture, or ethics, then the organization is creating visions that go beyond its bottom line. It will have more variety in its research and perspective and in the ideas it comes up with when addressing future threats or opportunities.

Cultural mindset shifts

A mindset that is transformed from myopic near-term thinking into more long-term proactive thinking is a good thing. If your work is able to change how the organization sees the future, responds to threats, and takes advantage of future opportunities, you are shifting the culture toward being more prepared and innovative.

Risk avoidance

Naturally, futures can allow you to identify threats early. It may be hard to measure that you've been proactive and responsible enough to avoid a threat early and survive, but once others are impacted by the same threat, you'll reap the reward of minimizing its impact on your organization.

Stakeholder or shareholder excitement

A well-crafted vision of the future will naturally excite stakeholders, especially if they had some part in crafting it. Having people who fund or lead an organization invested in the process and execution of the vision will be a huge success for you and the process and will allow you to continue to nurture futures within or for the organization. You'll have the funding, protection, and resources you need.

Cross-departmental collaboration

This isn't necessarily only a benefit of futures work. But since you need the organization to invest in the strategy, you may need as many resources as possible to execute on it. Bringing in other departments to help think, design, and execute could break down silos and encourage more cross-departmental collaboration, which in turn will provide multiple levels of support and excitement to help you execute.

Trend awareness

Setting up a trend repository or a process for people to begin paying attention to trends will start to change the way people are looking at the patterns
in the world. If you've set up a way for people to collect and discuss trends,
you can monitor engagement and how much the organization is responding to trends. Even if people are just moving from not paying attention to
trends to collecting them on a regular basis and using them to influence
other strategies in the organization, then you've made some headway in
teaching them to look around them and forward. Today there are many
trend analysis services (TrendWatching and Futures Platform, for example)
and AI-enabled tools (such as ChatGPT) that can assist in gathering trends.
As technology evolves, this will become a less manual activity, and we'll be
able to use (and trust) these automated systems to identify, analyze, and
synthesize numerous trends as they emerge in real time.

Shifts in strategies

If your futures strategy begins to influence an older strategy or develops
a new strategy for the organization, then you know it's been an effective
effort. Employing this process can have major impact on how the entire
organization thinks today and tomorrow. So if you're seeing large mountains being moved to accommodate your strategy, you'll know many are
listening and taking it seriously.

How Scenarios Can Inform and Transform Strategy

Jake Dunagan, Design Futures Director for Institute for the Future (IFTF)

We did a project once for a large automotive manufacturer, a luxury
automobile maker, and created several written scenarios, but I think the
real payoff, the real thing that moved the needle was that we created
some very detailed still images of city scenes for each of the scenarios.
You could click in and zoom in on each detail. There were probably 10
to 15 little narratives within different cities. One was New York, one was
Rio de Janeiro, one was Paris, and one was Rome. At a distance, you just
see a city and you get the sense of a "mood." In one scenario, the vibe
was misery, it was dreary and miserable, not a total collapse, not like *Mad
Max*, but it was just an unpleasant place to be. Another scenario was a
more bright green future, one that's super transformed and high tech.
And then another one was basically a collapsing civilization that was not

really holding together very well. But if you zoomed in, you could see little details. There's a clock with a temperature on it, there are fires on the hill, everything had a story tied to it. For a two-dimensional image, it allowed you a little bit of telescoping to focus in on the details, and they were meant to represent the scenarios.

What our clients did was take those and project them onto the wall whenever they met with their teams and their C-suite executives. And it had a real powerful effect. I know it did. This was the payoff, because it was within less than three months after we delivered that project that the automobile maker made a big announcement saying that they were going to move their carbon neutral strategy from 2050 to 2039. Which is a pretty big deal. It was also the year of the scenarios we designed. They're all based in 2039. So our clients told us that this was directly related to that work. And I think the urgency and the immediacy of an image, a representational image, broke through some kind of bureaucratic resistance that might happen or the sense that we can push that further. So, just turning up the heat a little bit on their thinking was useful.

Again, we weren't trying to take them in a different direction, but I think we communicated everything well in our work, and it was all backed by reams and reams of interviews and research. And we did a lot of internal interviews, and this was not something that we just threw at them as a template. It was really deeply researched and reflected their mindset too. So we were very much in tune with them because it was over a year that we were working with them and there were lots of meetings and calls and interviews. So we had a good sense of that. So all of those things were necessary to make that a successful project—the good work, the modification on the fly that they needed to be able to speak their minds, and also the communication techniques we used. And we were strategically positioned to be influential to their decision makers. And their policy changed, and it changed in the right direction.

INDICATOR MONITORING

Futures is never a one-and-done type of activity. Once a roadmap has been agreed upon, you may want to set up a cadence of checkpoints for your organization or teams to reassess the current and future landscape. *Indicators* are simply thermometers in the world that are dynamically changing or being impacted by various factors. Some are trends or are connected to trends, and some are specific stocks or products that you are monitoring to see whether they succeed or fail. They can be anything, including behavior patterns, qualitative opinions, or actual changes in political party popularity. Typically, the number of indicators and the frequency with which you check them, analyze them, and decide whether they impact your current projects or strategies are determined by your teams. Some teams may need more confidence in the beginning, so you may be watching certain trends or performance measurements weekly or monthly. As time progresses, you'll begin to decide which indicators need to be monitored more frequently, which are more nascent, and which can be checked every year or two. It all depends on what you want to measure and how you want to tie it to the success of your projects. Setting up stage gates or checkpoints periodically will allow you to do this assessment or reassessment to analyze what is working, not working, as expected, unexpected, or critical to address. If at any point something critical has changed, you may want to restart the process on certain projects or simply change a few tactics, goals, or visions.

Here are some questions you might ask at each checkpoint:

Are the trends we identified continuing to scale as we expected?

No matter how much quantitative and qualitative data you get about your trends, anything can happen. You can do a pretty good job of projecting, but you never want to stop observing these patterns as you move forward. At every strategic checkpoint or reassessment moment, you'll want to review the mega trends but also, more importantly, the weak signals. Are the macro and micro trends scaling up or down as projected? And if so, are you prepared? Does the strategy compensate for the changes and implications they are creating? If yes, great; if not, it might be time to make some changes in your plan. You may need to do another trend analysis, implication, and scenario exercise to consider how changes in trend trajectory could ultimately affect your current and future workstreams.

Are there any new innovations, technologies, or disruptors?

You may have done a good job of identifying emerging innovations and technologies, but new ones could pop up at any time. There will certainly be things that you didn't see. How do we factor those into our current strategy? Does anything need to change dramatically, or do we just tweak a few things here and there? Or do we need to make a small pivot to take advantage or avoid some new disruption we are seeing? New innovations in tech or processes can become major disruptors and can appear out of nowhere. If you are monitoring the industry, you might see these weak signals early enough to respond to them. If you notice something that could have a high potential to threaten or enable your business or product, call attention to it, analyze its drivers and trajectory, and decide whether it's something you need to do deeper work to investigate or incorporate into a new futures strategy.

Are there any anomalies or black swans?

Investopedia (*https://oreil.ly/9paP2*) defines a black swan as "an unpredictable event that is beyond what is normally expected of a situation and has potentially severe consequences. Black swan events are characterized by their extreme rarity, severe impact, and the widespread insistence they were obvious in hindsight." Some believe the coronavirus pandemic was a black swan event because it completely blindsided many of us. When such an event occurs, you naturally want to quickly reassess your strategy and identify whether it has any impact. Similarly, any anomaly, such as a sudden change in the direction of a trend or technology, could be disruptive or may require you to rethink your strategy, especially if you were investing on it to build a business. For example, nonfungible tokens (NFTs) are block-chain-enabled digital art pieces that rapidly grew in popularity around 2020 and saw a spike in consumption. However, as reported in a September 2023 *Guardian* article (*https://oreil.ly/4bg9Y*), an investigation by the crypto gambling platform dappGambl in which a team reviewed data from NFT Scan and CoinMarketCap found that "69,795 out of 73,257 NFT collections have a market cap of 0 Ether, leaving 95% of those holding NFT collections—or 23 million people—with worthless investments."

Did we achieve our near-term goals and expectations?

Every step on your roadmap should be a measurable one. Make sure to assess your near-term successes, log them, communicate them, celebrate them, and connect them back to the greater strategic roadmap that influenced them. This again reinforces the importance of the futures work and the connection between long-term and near-term successes.

Did the market appear?

Even though you may have a strong and well-informed strategy based on rigorous research, an emerging market or need may or may not appear. Thus it's important to continue to monitor your trends and other indicators, especially if you are taking big leaps of faith or risks with a strategic initiative. If you are seeking to pioneer into a new market and gain new customers, you simply may not see the emergence or growth of the market you expected. Sometimes products enter what we call a *hype cycle*, a term coined by the American technological research and consulting firm Gartner. Gartner's Hype Cycle (Figure 11-11) is a graphic representation of the maturity and adoption of technologies and applications and how they are potentially relevant to solving real business problems and exploiting new opportunities. Some new technologies that suddenly gain a lot of media attention or grow into mega trends quickly go through a phase of "hype" then disappear (like NFTs, hoverboards, or fads) and enter the "trough of disillusionment," where society doesn't deem them valuable anymore (the market never appeared, or it disappeared). Then maybe something changes, technology improves, and you begin to see them again, and suddenly they are useful, society becomes "enlightened," and they flow into the "plateau of productivity," where they become popular products again.

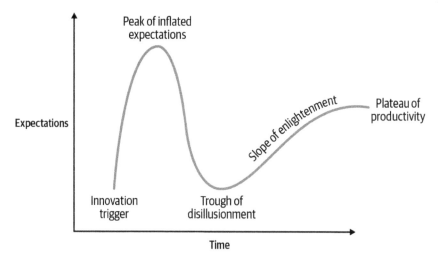

Figure 11-11. Gartner's Hype Cycle

How does the organization feel about the plan?

Every so often, it is nice to know if the team or organization is still excited and on board with the greater strategy. If at any point you discover skepticism or a loss of excitement, you may want to initiate a conversation or

survey (informal or formal) to understand why. While you can't always keep everyone happy, small disappointments, disagreements, or negative emotions toward the plan can appear or stack up, spiral out of control, and influence major stakeholders, which could put your plan at risk of being shut down entirely.

Have new and exciting developments emerged as a result of our efforts?

Much like how black swans can surprise us and have severely damaging effects on us or our plan, positive surprises and developments can emerge also. Has anything new, constructive, or exciting happened due to the implementation of your plan that you didn't project? Keeping an eye on these developments can also bolster the success of the strategy. If another department begins working on something similar or was inspired by your work and began developing it further in a different business unit, it's a positive sign you are making an impact. Whether you see these signals in your company or in the market, every win counts.

Were we completely wrong about something?

If everyone has an optimistic futures mindset, then failures won't be so disappointing or damaging. Yes, it is possible to be wrong or inaccurate. That's why we don't use the word *predict*. What we do when we discover we were wrong is what matters. If a trend or pattern didn't scale the way we planned, hopefully we at least saw that coming and have some contingency to course correct (a Futures Wheel is always good for assessing potential risks). If not, then it might be time to assess how a failure or an inaccurate projection might impact you. It's important that everyone knows it's OK to be wrong and that they keep a positive attitude and look at it as an opportunity to improve or pivot toward a positive outcome.

Should we start over?

The future can bring many surprises. And sometimes those surprises come with such massive implications that you may need to start the process over again. This doesn't mean you did it wrong the first time or that you failed! But maybe your research had a blind spot (it happens) or some weak signals accelerated in a different direction than expected, and you weren't entirely prepared. That's OK. In fact, it's a healthy opportunity to start from scratch every so often, especially if you have learned more methods, captured more trends, and learned a lot since the beginning or last strategic reassessment. A reassessment checkpoint can include doing the whole process over again to make sure you did it right or to reinspire stakeholders toward a new vision or variation of the old vision. You might even want to evolve or refine the process. Maybe you've created your own new methods you want to try out, or you want to do it all over again in collaboration with a new partner or additional voices in the room who may have been absent before. Starting over is a great way to revitalize your process, get more practice, and reinvigorate your campaign for the future vision. Keep in mind that "starting over" is not actually starting over. You are just reassessing based on new conditions and climates. Again, having a futures mindset means that you are able to accept the multitude of possibilities (good and bad) the future holds. Having another workshop to assess a new set of trends with the expectation that it could in fact change your entire long-term strategy is actually a common practice. We see this all the time when new technologies or trends appear and companies scramble to adopt or compete so they don't become irrelevant.

The intent behind monitoring indicators is to determine whether you are still directionally correct, but you need to install this process and program these checkpoints to decide how and when you will do these assessments and reassessments. Maybe your organization is still very new to the process—do you need to do these check-ins more regularly? Quarterly? Yearly? At each checkpoint, try to have a set of questions and checks you want to perform regularly.

Implementation and Delivery

Executing a strategy, whether it was delivered by a consultant or by an internal team, can be tricky. No matter how much work you do planning a roadmap, allocating resources, and securing the budget, many factors can impede the delivery and implementation of that strategy:

Roles may change.
    Whether you're working in-house or as a consultant, you probably have an internal team that has been helping with or is responsible for the futures project and strategy. In some cases, after the work is done, roles may shift or change. The leader or manager who initially sanctioned your work may quit, get promoted, or move laterally into another part of the organization. Without this sponsorship, ownership, and leader to continue to oversee and shepherd the process, the work could dissipate and be forgotten.

Priorities shift.

You may come out of a strategy meeting with a lot of excitement to execute. The organization could seem completely committed. However, as the future does, things can change. Priorities within the organization could shift. A dramatic change in events could occur for any number of reasons: a new CEO, rising competition, new emerging technologies, shifts in available funding. The leaders who were initially assigned to help implement the project could be brought over to other strategies, deprioritizing the futures work.

Funding is lost.

A harsh reality we all face in business is that businesses run on revenue or funding of some kind. When money is not available, operations are impacted. You could discover that the funding initially allocated toward implementation gets shifted around or is lost entirely. A long-term strategy with near-term activities requires the financial dedication to make it work. Without it, you're dead in the water.

Another strategy is created.

If you've ever been on the giving or receiving end of consultation work, you'll know that sometimes a consultancy is brought in to hand over a strategic plan, and sometimes the strategy changes after the consultants leave. The organization may take the initial plan and modify it. This isn't necessarily wrong—an organization needs to refine the strategy to work for them; however, without leadership and guidance, the initial strategy could be completely scrapped and replaced by a new one that completely ignores some of the work you invested in creating the original plan.

If you are part of an implementation team, you'll have the opportunity to see the future unfold and assist in its development. However, sometimes it is completely out of our hands, and we might need to let go, because we can't control the gravitational forces that might be pulling an organization in other directions. The best thing to do is to be aware of potential situations that could occur and set yourself up for success as early as possible—always practice futures, and always look ahead.

In Closing

In my career path from designer to director to strategist, I've found that it's both natural and important for designers to begin to learn what *strategy* means and how they can participate in that conversation effectively. Strategy should not be a foreign or intimidating function. In fact, it is a mindset and set of principles that I've been trying to reiterate through numerous analogies and examples throughout this book. Designers are, in fact, strategists by nature. We look around us, behind us, and ahead of us to determine what are the best solutions we can craft that solve problems and provide a positive impact to our users and customers, society, and the world at large. Being a strategic designer just means that we have a set of tools that allows us to have foresight and vision and that we know how to plan to execute that vision. We're able to look ahead and beyond what others are able to at this moment, make informed decisions about what we believe the world needs, and execute on those ideas through our craft of research, imagination, and focus on the architecture of experiences.

If you are able to lead and design the strategy for your future vision, product, service, or initiative, remember to:

- Think about your audience and who you'll invite to the planning session.
- Prepare everyone appropriately for what the workshop entails and the expected outcomes (what we are doing, what we expect to accomplish).
- Select your activities, timeline, and agenda.
- Determine how to make the long-term strategy actionable through the setting of near-, medium-, and long-term goals that align around the vision.
- Consider a variety of ways to measure success along the way, from immediate ROI on projects to long-term cultural, market, and customer successes.
- Support the implementation of your futures vision and strategy if you can; oversee, monitor, and guide your client or team through the first six months or year to make sure things go smoothly and to sustain the excitement and futures culture of the organization.

In the last few decades, as design has become a more important and integral function of business, we cherish those moments when we are "invited to the table." Looking back on how, when, and where I've had the opportunity to practice strategy, I've always felt that my education was lacking an understanding of the cultural and political nuances of how to make strategy work at every level of an organization. The most important learning for me is that strategy is more than a planning activity. It includes having a vision and a commitment to that vision as well as expertise and experience in project management, people management, sales, advocacy, evangelization, change management, and execution. It also demands a sense of rigor, an ability to comprehend the complexities of how people and organizations operate, and the charisma and candor to manage up, down, and laterally to get things done. Futures Thinking is naturally a strategic exercise and can be seen as complementary to other strategic practices, so it should be treated delicately as you begin to integrate it into your organization (whether from the inside or the outside). In the next chapter, I'll discuss some of these entry points and share some stories about how others have made those first steps, successfully or unsuccessfully.

of the cultural and political nuances of how to make strategy work at every level of an organization. The most important learning for me is that strategy is more than a planning activity. It includes having a vision and a commitment to that vision as well as expertise and experience in project management, people management, sales, advocacy, evangelization, change management, and execution. It also demands a sense of rigor, an ability to comprehend the complexities of how people and organizations operate, and the charisma and candor to manage up, down, and laterally to get things done. Futures Thinking is naturally a strategic exercise and can be seen as complementary to other strategic practices, so it should be treated delicately as you begin to integrate it into your organization (whether from the inside or the outside). In the next chapter, I'll discuss some of these entry points and share some stories about how others have made those first steps, successfully or unsuccessfully.

Integrating Futures into Your Organization

Once you feel confident about being a practitioner, you may want to start thinking about how to begin integrating Futures Thinking more formally into your team or organization. In this chapter, I'll discuss a few different points of entry. Whether you are an in-house employee (from the inside) or working for a consultancy or agency (from the outside), there are a few tactics and strategies that are similar to what we've already discussed in previous chapters about preparing your audience for the mindset (check out Chapter 3 and Chapter 4 for more on that). But since every organization is different, there will always be cultural and operational nuances that you'll need to consider to make your campaign work. Because in the end it *is* a campaign and an initiative, and it will require dedication, commitment, and, above all, patience to integrate your vision. It can be a lonely journey at first, but with proper research, planning, and experimentation, it's definitely possible to make futures work in any team or organization. It just may take longer in some cases than in others.

From the Inside

As mentioned in Chapter 3 with regard to understanding your culture and audience, it's important that before trying to advocate for anything new, you are informed and mindful of the environment within which you are working. Even in the best-case scenarios, where you are hired into a team that is already practicing futures, you should still be aware of the politics, context, and current approaches so that you don't inadvertently disturb or threaten any process or people in the organization today. Bringing anything new into a workplace can be sensitive and risky, so you'll want to make sure you have the support to do so.

Here are just a few common situations you might find yourself in as an in-house employee:

- You'll be the first to introduce futures to your team or organization.
- You've established some support within your group or team and want to introduce futures to other teams within the organization.
- You're hired to lead a team or to be part of a team that is or will be intentionally practicing futures.
- A strategic function or business unit already exists that may or may not use futures, and you would like to collaborate or introduce it to them.

YOU'LL BE THE FIRST TO INTRODUCE FUTURES TO YOUR TEAM OR ORGANIZATION

You might be on a team where no one has ever heard of Futures Thinking. Or you may have encountered a few people who have heard of it to some degree but don't practice it. In this situation, futures is a new process for everyone, and you'll need to find the best way to introduce it so that you can gain the support system you need. If this sounds familiar, it's because we've already talked about some tactics for getting to know your culture and environment, and I suggest using very similar investigative models (such as a diagnostic) to get the lay of the land. But there are a few other tactics you might employ:

- Deliver a presentation
- Find supporters and champions
- Invite an external speaker
- Distribute or post inspiration internally
- Have a lunch meeting (brown-bag it)
- Conduct a workshop
- Advocate for a project
- Use a method on a current project
- The Trojan horse

Deliver a presentation

A presentation is always a great way to introduce people to the process. You can include exciting and provocative examples,[1] expose the theory and methods, and give people a taste of what the outputs could look like. This is a chance not only to educate but to inspire. As the potential leader of the effort, you can also set yourself up as the expert. People who attend will then know you are the one to talk to if they have questions, and you can begin solidifying your role even if it doesn't exist yet.

Find supporters and champions

Do a little digging around and see who might be interested in helping you introduce futures. There may be one or two designers who have heard of it but never used it. There could be an executive leader who loves it but has never found a chance to implement it. Even just having one person on your side is great. But ultimately it can take a village to implement futures, so finding as much support as possible early on is important. Once you've found those people who can help champion your efforts, you can strategize around what is the best approach for maximum impact.

Introducing Futures at IBM

In 2017, Michael Kenney and Gabriella Campagna Lanning, who at the time were strategists and researchers at IBM, teamed up to begin introducing Speculative Design and futures to the company. What started out as a casual conversation and sharing of interests would eventually lead to developing a toolset that would be implemented into the Enterprise Design Thinking training catalog that IBM used to train all of its employees, and which is also publicly available. Michael Kenney recounted the process in a conversation with me:

> Gabbi and I met on our first day at IBM. We came into a cohort and then quickly jelled around some ideas, and just became friends both inside and outside of work. We had kind of already started to do some extracurricular things: we filed a patent

---

1 See "Show Relevant and Appropriate Examples" on page 84 for help in determining what examples to include in a presentation.

together within our first couple of weeks and had a couple of educational projects and other things within the design community. We were sitting in the kitchen one day, and I asked, "Are you familiar with Speculative Design and future studies?" And she was, and I said, "Wouldn't it be awesome to have a job title of futurist?" And then we started asking how we could do this at IBM. There are no futurists really at IBM. They have some people in IBM research that are working on cutting-edge technology, but nobody that really sits in a futurist role. So how do we put a program like that together? So we did some brainstorming. One day, I was looking at my bookshelf, and I remembered that I had this book that was called **Twelve Tomorrows.** *It was an anthology that they would publish that would take these leading science fiction authors, and then they would give them a new emerging technology, like conversational AI, and they would pair the emerging technology with a well-known sci-fi author and ask them to write short stories. At IBM, we had this thing called* **Five and Five** *where our IBM Research Division would actually put together five technologies that they think would change the world in five years and then they would publish that at our annual conference. At the time, IBM Design had a real culture of making, and we had maker spaces and these grassroots cultural initiatives. So some designer would say I want to be a DJ and then IBM Radio spun up. That eventually became a global network of people who could log in and listen to IBM Radio, and people from other countries around the world who are DJs. We also had a global network of maker spaces where people had screen printing supplies, 3D printers, and other rapid prototyping stuff.*

We also had an internal magazine that the designers had put together, and it was super high production value. They would pair, like, illustrators with writers internally, and they would have a theme and would publish it quarterly. And so it turned out that Gabbi had worked with the guy who had originally started this magazine and had built it up over the course of four or five years. So we had this **Five and Five** *publication and this magazine that's distributed to studios all over the world, and they're always looking for content. I was in the world of Content Strategy, and*

I knew a bunch of writers who were also amateur science fiction authors. And so I said, why don't we mash all these things up— the sci-fi and the Five and Five *predictions—and talk to someone about publishing these in his magazine, and everything just came together.*

So we announced it on Slack and said, hey, we have a science fiction project. Are there people interested in contributing? And we received submissions from people all around the world. We paired them up with one of the Five and Five *concepts that were coming out. And so we published these short science fiction stories and called it* Sci-Fi and Five.

At the time there was a really big push for designers to become part of IBM's patent community so that we could actually achieve levels of inventorship, similar to what the engineering and business teams were doing. IBM places a lot of value on IP. Every year, they're miles ahead of everybody in terms of patents filed. So that was a success metric that we could report on, an ROI on investing in designers, how they could contribute back to the bottom line.

And so what Gabbi and I did was really push the writers to really detail out the worlds they were building, and we tried to implant artifacts and processes in that world. Not really telling them what our intention was, but afterwards what we did was we audited all of the stories and said these are all the individual nuggets of potentially patentable ideas or pieces of technology or something that could fit into our product roadmap. Fast-forward a month later, and we had managed to get some time on our highest design executive's calendar. We put together a presentation that showed that we had 45 ideas of individual IP that we were able to generate in less than 6 weeks' time from people all over the world.

We gave the SVP a rundown of all the different business units that this could apply to and then showed him the glossy output and that we had actually built a community. But what we were most interested in was the idea that those ideas had been attached to a story where those artifacts were examined in the lens of their impact and technologies. If we were to fast-forward a number of years, what is the potential for cultural, social, environmental impact? IBM Research was just producing this publication called **Five and Five** and saying this is going to be a great technology that's going to be a big part of our world in five years, but they weren't actually doing the type of Speculative Design that I think people were most interested in understanding—the impact or some sort of impact measurement. We were saying, here's the technology, here's how it manifested and influenced the world around it, and here's the potential impact. So what we were really trying to do with the Trojan horse was disrupting the culture of building technology. If we were going to have a sustainable culture, it needed to be around a program of Speculative Design and futuring. And the SVP was interested in the list of IP, and his eyes opened up and said, this is amazing! Eventually, because of the volume of patents that we had, we were assigned a dedicated IP attorney who took our spreadsheet and started going through each idea one by one with a whole IP team.

After that, we said we could try to spin up a futuring organization or an innovation hub, but it still felt fairly exclusive and maybe it wouldn't have a lot of visibility. At the time, IBM had an enterprise design thinking campaign that they were scaling; they wanted to turn every IBMer into a trained design thinker through this platform, which had some methods, toolkits, and this philosophy of design thinking. And that was almost like a corporate mandate coming from our old CEO. So that's when we said we should just ride that train. But there had never been any kind of individual contributors who had created new toolkits for that. So we said, we'll put it into a toolkit (https://oreil.ly/qNSLM) where anybody from any education or background can jump into this and start to create these kinds of speculative fiction stories, and then we'll show them how to go through and mine the IP from there.

It was interesting because it forced the design organization to actually think about how an individual contributor contributes to design thinking, which again was a catalyst for change for them because we had all these expert designers we've skilled up, then why can't we have a more inclusive culture about people contributing new methods and ideas to continue to evolve our programs so we don't get stale. So that's what we did, we published that toolkit, and then once that toolkit was out there, it started to get a lot of attention from executives in different areas. Then once COVID hit, all of a sudden there was an innovation Task Force tasked with figuring out how do we get a bunch of good ideas about how to deal with COVID using IBM technology, and we started getting invited to those meetings and actually started to build up some momentum around having resources dedicated to those activities.

Invite an external speaker

Sometimes it's better to have someone from the outside come in and talk to your team, rather than you doing it. This relieves you of some pressure or criticism that could occur if it doesn't go well. It's also a way to silently test the waters to see if there is excitement or appetite. You could simply frame it as inviting an inspirational speaker and not reveal that you have an intention to integrate futures into the team or organization. Some teams may not grasp certain principles well or may not have an interest in futures work. But try to work closely with the speaker to make sure the content is tailored to your team's domain, understanding, or needs. Help them choose the right examples and rhetoric to use, understand who will be in the audience, and decide on what outcomes you expect from this presentation. Do you want to use the presentation merely to stimulate discussion about the topic? Or do you want to use it as a marketing tactic to invest in deeper training? These questions will help you decide on the appropriate kind of language and content for them to use.

Distribute or post inspiration internally

Whether you like it or not, this is a campaign, and within campaigns, propaganda or marketing tools (images, videos, articles, blogs) are used to gain interest in your topic.[2] These can be anything from news articles and projects to trends and signals. Try to find out what are some of the popular communication channels people are using. Is there a periodic newsletter or email, and if so, is there a group that is responsible for it that you may want to partner with? Is there an internal communications platform such as Slack, a wiki, or an intranet? Find something free, cheap, and easy to start with. If you begin to receive inquiries, you can decide what the next step is to provide more information or inspiration about the work.

Have a lunch meeting (brown-bag it)

A lunch meeting is a great, casual way to introduce fun and extracurricular topics. A "brown-bag meeting" is just a meeting that you bring your lunch to. This should be a low-pressure, zero-dependencies, zero-commitment meeting at which you can simply introduce a topic and have a short discussion about it. You may not want to reveal your intentions behind the meeting at first to test the waters, but again it's a nice way to see who responds affirmatively to futures.

Conduct a workshop

A great intro into futures is to throw people right into an exercise. Introducing this idea could take some negotiating with your manager, and it might mean finding the right time or project to suggest it for. An easy way to start a workshop is to do it "off the clock," such as during lunch, after work hours, or during downtime when you're waiting for a project. Once you find that time slot and opportunity, think about which method will be easiest to understand and facilitate. Whether you have great results or not, a low-stakes ideation session is always a quick way to get people to exercise a part of their imagination that they might rarely use in their daily work. Consider choosing a topic outside their domain (e.g., food or outer space) to remove them from their industry thinking.

---

2 The use of the word *propaganda* is not meant to be divisive or fake. Propaganda is a general term used for marketing efforts to gain publicity or expose content in a variety of formats. Historically, propaganda was used in posters or flyers to sway public opinion for political purposes. Inspiring people about the future can be a political effort in the sense that you need to gain interest, followers, and advocates for the use of futures and a mindset of anticipatory thinking. So whatever you call it, propaganda or marketing, the use of artifacts to educate and inspire is one of the tools you may have to rely on.

The workshop doesn't have to be too long either. You could explore just one method or one trend, or just have a general brainstorm about the future without using any methodologies at all—whatever works for your audience in the time you have.

Advocate for a project

You might find that there is a project in your portfolio or pipeline that could really benefit from a futures approach. Find a way to advocate working on it. Again, if you've thoroughly diagnosed your team, culture, and projects, you'll know who might be able to open the right door for you. You may even have to spend a year or so just building up trust and recognition with your team before you feel confident enough to suggest such a topic. It could take a few attempts and a variety of tactics and suggestions to get a project started. You may need to determine how to make the project valuable to the company or clients. What are the goals you hope to accomplish? How will a new project like this impact the business or customers? How will you measure its success?

Use a method on a current project

You may want to try smuggling a method or two into a workshop. You don't have to necessarily say you are using a "futures" exercise; you can just call it an alternative "method" or "tool" for understanding the implications of a feature or product. Once you have a success, you can suggest more, or you can fully expose the process when you are ready. Disguising or relating a method as something that people already use and know can be helpful when you really just want to give people a small taste of what futures approaches are like.

The Trojan Horse

Sometimes, when introducing a new process, it helps to hide your intentions by disguising the process within another process or system that has already been adopted or validated. We sometimes call this a "Trojan Horse" in reference to the story of how the Greeks ended their 10-year siege of the city of Troy. The Greeks built a giant wooden horse (Figure 12-1) and left it outside the Trojans' walled compound. Believing it to be a peace offering, the Trojans brought it into the compound. That night the Greek forces hidden inside the horse surprised and conquered the city of Troy. While we don't necessarily want to *surprise* anyone, the Trojan Horse is a model we sometimes use to introduce something that could be sensitive or that might be rejected immediately. You may want to engage in a traditional design process or ideation session before casually suggesting another method (like the Futures Wheel) to arrive at an outcome. Once people are comfortable with that, you can try introducing other methods, thereby continuing to expand the conversation.

Figure 12-1. The Trojan Horse

Perhaps you're involved in an ideation for a product design. Introducing some of the futures ideation methods discussed in Chapter 9 could help your team come up with new and novel ideas and allow them to reframe how they are thinking about the design. Again, you may not want to say this is from a futures toolkit, but you could begin to open the conversation to the other tools you have at your disposal.

Another approach within a strategic conversation might be to assist a project or client in getting to the near-term goal—something you are trying to build within the next 6 to 12 months—and then ask questions about what happens beyond that. Some projects have hard deadlines directly defined by the potential scope and effort of the work necessary to build the product or service. Some organizations aren't interested in looking beyond that point unless they've already moved the needle or made some progress that they can strategize around. Introducing a longer-term strategy, or at least a conversation, could open a door to discussing future implications or scenarios. In fact, even at one year out, you could introduce the possibility of alternative scenarios. How might this product fail if a competitor advances with a technology first? What if the economic, technological, or political landscape changes drastically in one year? What do we do then? Simply suggesting alternative futures is the beginning to a conversation that could employ the toolkit. You wouldn't necessarily need to expose that there is a whole methodology for thinking about the future yet, but if you are working with stakeholders who are receptive to the notion of talking about alternative long-term scenarios, you have a door open.

Using a Trojan Horse is not meant to be a divisive or deceptive way of working but rather an alternative shortcut to showing others the capabilities of the tools futures has to offer. Use it with discretion, and try to measure what is or isn't working. This information can be valuable in telling you how easy or difficult it might be to try to propagate the process internally or to other parts of the organization.

YOU'VE ESTABLISHED SOME SUPPORT WITHIN YOUR GROUP OR TEAM AND WANT TO INTRODUCE FUTURES TO OTHER GROUPS WITHIN THE ORGANIZATION

If you've been successful in implementing the process within your team, whether through a formal project, training, or a mock project, then you've already done a lot of the hard work in establishing a foundational support system. Hopefully you have some evidence from the team's work that shows what the process looks like and how long it took, method outputs, the problem you are solving for, and an inspirational vision that makes sense for the company's ethos. Before showing others, make sure you have some documents that can explain what you did clearly and concisely for any newcomers. You may even have different documents for different types of audiences. The pitch deck for a CEO will be much shorter, with fewer words, more pictures, and maybe some data points. The pitch deck for a strategist or product manager could be a little more robust and have very specific examples, visualization, and descriptive themes, with pictures of people working in workshops (or of digital whiteboards). Think about your audience, because once you start exposing others to the work, it will be like starting at square one again—you'll have to do a bit of priming to get people interested and understanding the value.

Some organizations may already have a strategy function. In some cases, the word *strategy* could be triggering for those who are experts in business management strategy. You may need to steer clear of words that could make people feel territorial about their practice. Futures should be a collaborative affair, and strategists would be some of the best allies, but there can be a world in which you operate independently of them if they do not see the value in how you are practicing strategy through the use of futures methods.

Introducing Futures at Twitch

As a designer in a large tech corporation, introducing speculative design can be challenging when there is little knowledge or awareness. Shihan Zhang, Design Futurist and Design Lead at Twitch, has been practicing Speculative Design with her studio Alter+ (Alter Plus), but until recently she has kept that work separate from her day job. This is common among those who are interested and passionate about futures work. Without support internally (funding, time, and projects), bringing the practice into an organization can be difficult, but it's not impossible. As an experiment, and to promote the practice on one of Twitch's internal products, she created a small exhibition in one of the kitchens at the company's headquarters in San Francisco (Figure 12-2). The installation included mixed-media artifacts—videos and interface prototypes that allowed people to stop and experience their future product ideas on the way to get coffee or food. Many people interacted with the installation, including the CEO and other leaders. The project impacted so many people during the time it was installed that it inspired a shift in one of the product teams' strategies for the following year.

Figure 12-2. Installation at Twitch headquarters in San Francisco

YOU'RE HIRED TO LEAD A TEAM OR TO BE PART OF A TEAM THAT IS OR WILL BE INTENTIONALLY PRACTICING FUTURES

This is probably one of the best-case scenarios. You'll be very fortunate if you've found an organization that already understands the value of Futures Thinking and is well on its way to implementing it (or at least beginning its journey). However, even with proper funding, political support, and resources, there may still be some challenges ahead. Get yourself acclimated to how things work today. How is futures or foresight being applied? Who is leading it? Who is trained or not trained? What are the projects that are foresight enabled? What are the success measurements (KPIs) being used? What challenges has the team found over time? What opportunities is the team shooting for in the near or distant future? How are initiatives created and qualified today? Who signs off on strategic projects? Depending on where you are inserted, you could have an easy and exciting time integrating yourself into a family of practitioners—or there might still be some work to do. Luckily, in this situation, you may already have some foundation to work with, and so you won't have to do as much advocacy or evangelization of the process.

Building Foresight Capabilities at Autodesk

In 2017, Radha Mistry, Americas Region Foresight Leader at Arup, joined Autodesk and eventually formed its first foresight team and practice:

> When I first started, there wasn't a foresight practice. And I came into the Immersive Experiences group, so I didn't even have any role in strategy or corporate strategy or any kind of long-term research. And when the Immersive Experiences group was dissolved, I got pulled into the corporate strategy team. And at the time, we were nested under the CTO's organization, and that was linked very closely with the research organization—the group that is known for being the far-out futures thinkers—but oftentimes there was a cognitive dissonance between what research was doing and what the rest of the organization was grappling with, and as someone who was really new to the organization, that was just blatant to me. When I was starting to talk to people, everyone had a different concept of what "future" meant to them. It was a different time horizon and a different language. They were looking at different tools and methodologies and realized that there wasn't a consistent practice or uniform point of view on what we think is changing in the world and what we believe we should be paying attention to.
>
> What I essentially did was create a foresight audit. Me and my research partner went to the different teams and asked them, "How do you think about the future, what kinds of tools and methods do you subscribe to, even if they're not foresight-oriented (because nobody really knew what foresight even was), and what value do you see in thinking about the future?" And we compiled that into a report (because that's the best way to grab people's attention inside corporate sometimes) and then turned it into an abridged presentation and just started socializing that across the company, primarily with executive leadership. And our thesis was that we can better capitalize on change and come from this place of agency and ownership around the future if we have one common language and a practice for how we address

the future. Wouldn't that be wonderful to not feel like you have to resign to the future happening to you? So to be able to better bridge between what research is doing on a 20-year horizon versus what our customers are grappling with on a 2-year time horizon, we were asking them to invest in robotics. But they were saying, "Our building codes aren't even changing now, how do you expect us to adopt new technology when the policy isn't even there? We're still taking on the same liability and have 2% profit margin." So it was a massive mess of stuff, and the serendipitous thing that happened at the time was the nature of who owns strategy inside of Autodesk was shifting, so there wasn't really a need for corporate strategy at that time; that was an evolving thing, and strategy was being pulled into each of the industry groups.

So architecture-engineering-construction and then product design and manufacturing wanted to own their strategy, as opposed to corporate strategy handling everything. And so we were in this moment where my VP at the time, John Pittman, was starting to have conversations with our CTO about what we become. And I had been going around saying that foresight can fix everything! Which of course it can't, but there are really good problems that foresight can help with.

And so, with the support of our CTO and later our CEO, we ended up rebranding the corporate strategy team into strategic foresight, and part of that came from tapping into the fact that he really loved science fiction, And he saw the value in science fiction authors being able to amplify weak signals of change and help people get ahead of that change in meaningful ways. So we leveraged that. And before we officially became a foresight team, the first engagement that I ran was with our executive leadership team because our CEO wanted his executive leadership team to become fluent in foresight, and eventually we had annual engagements with them.

A STRATEGIC FUNCTION OR BUSINESS UNIT ALREADY EXISTS THAT MAY OR MAY NOT USE FUTURES, AND YOU WOULD LIKE TO COLLABORATE OR INTRODUCE IT TO THEM

If you discover that there is another team, function, or business unit or other strategist roles that are already doing long-term thinking and planning, these can be your allies, competitors, or skeptics. Naturally you want to reduce the friction between you and other practitioners, so try to find a way to speak with them, discover their process (and personalities), and see if there is an interest in learning about futures or collaborating with you on a project (real or for practice). In your diagnostic, you may have discovered these people, processes, or groups already. Since every organization's culture is different, try to employ some of the tactics discussed in Chapter 3 and Chapter 4 for understanding your audience and priming them for the futures mindset and process. This could be the entryway to strong partnerships and mutual learning. Again, be mindful of what examples you show them and how you explain the value. Some strategists might require quantitative success metrics and evidence that your process is similar, complementary, or imperative to their work. This advice goes for any other function in an organization that could be useful in advocating or practicing futures or integrating it into your workplace. Take the time to understand who you need and how to be their allies. And if you do face conflicts, choose your battles wisely, leverage your champions and support network, or find another strategy to work with or without them to move forward.

Operating from inside an organization gives you many advantages for integrating futures. You have more time to understand deeply how the organization works and what the best paths to success might be. But no matter where you are starting from, politics, funding, and agendas can always get in the way. Try to have a long-term strategic vision for how you want to begin these conversations, and employ as many tactics as possible to move forward. You might even take inspiration directly from others who have done something similar (in fact, partner with them if you can). Whatever the situation, try not to get discouraged if futures doesn't catch on quickly. It can be a long game you have to play, but it is well worth the wait if you can manage to transform the way an entire organization thinks about the future.

From the Outside

If you are working as an external party, such as a consultant,[3] you will have some freedom and privileges to be an external voice to an organization, which can work for you or against you. Similarly, there are many ways to begin introducing futures, but you'll also need to be mindful of the politics, people, and environment you are working with. That said, there are also many opportunities that come with working as an external advisor. Some you'll want to take full advantage of, and some may require a little more care, dedication, and effort for you to capitalize on them. Some of the tactics we've already looked at in this chapter can also apply to operating as a consultant, but here are a few other situations that are specific to what could happen when advising from the outside:

- You are hired specifically to use futures on a project.
- You're hired to train an organization or team.
- An organization is curious about futures and is asking you to introduce the topic.

YOU ARE HIRED SPECIFICALLY TO USE FUTURES ON A PROJECT

If you are a consultant and are hired specifically to provide expertise on Futures Thinking on a client project, then you already have one foot in the door. But similar to when you are hired internally to lead a futures team or project, you may have it easy to a degree, but some work might still need to be done to nurture the practice. Every organization is different, and its level of literacy could be at any stage. But if the hiring organization or client already has an investment and interest in the process, then that is a great way to start. You'll still want to employ the same tactics mentioned in Chapters 3 and 4 for understanding the culture and parameters you are walking into, to ensure that it can be a safe and healthy journey for all. Here are some caveats you might want to be prepared for:

---

3 Consultants come in all shapes and sizes, and depending on the agency or firm, you could be operating from any role or have several responsibilities, from being a project lead/manager to designing and prototyping. This section speaks to those who are working as an external party and advising or doing the work for a client as a consultant or an external worker. The intent is to show some of the (political, operational, cultural) nuances of working externally as compared to being an in-house internal employee. But the advice here can also be applied as an in-house worker, depending on the configuration of the organization. Your "client" could be your boss, another team or business unit, or a client of your company.

- Consider (based on the client's current futures literacy) how much you need to update or prepare them for the process.

- Do you need to omit, repackage, or use different terminology for them to understand the fundamentals of futures?

- What happens after this project? How and where will the organization apply it? How will it execute on the strategic outcomes?

- Is there an opportunity to install a dedicated futures team within the organization to oversee training and trend scanning and support the strategic initiatives?

Even if there is an investment, you may find friction or barriers along the way. Use your diagnostic tactics to identify these potential barriers early and create a list (mentally, in writing, or as part of a discussion with your team or client) to determine the course of action should certain scenarios impede your work. And if you have time, try not to rush the process, because that could create compromises across many areas necessary to set a healthy foundation for success later.

YOU'RE HIRED TO TRAIN AN ORGANIZATION OR TEAM

If you are confident enough in your knowledge and/or experience using futures, you may want to begin training your client. As a beginner, it's always good to practice or experiment on real or mock projects just to get a sense of how to facilitate and to determine how you want to train others to use it. You are welcome to use others' examples or training programs (if you have permission or they are open source) and think about how your program, including your style of teaching, will be most effective. Organizing small workshops is a great start and could lead to more requests to teach or even to lead a project. But training people how to use futures has its own advantages and challenges. If you are hired as an external trainer, you may not have much commitment beyond showing people how to use the methodologies. So the job may be easier, but the challenge will be HOW to train them so it's useful for your client. You also may be given only a limited amount of time to train them.

It's always a good idea to have an understanding of how much training you can provide in the time you have been given and what is necessary or unnecessary to teach in these first stages of building your client's literacy and capabilities. A Futures 101 course may not need to include every single method

in the playbook.[4] You may want to tailor the content based on the type of work the client will be applying it to in the end. You may even want to use the training as a way for them to practice the process on a real project, making it both a consulting and training effort simultaneously. The experience you provide can also make or break this effort. If you are not careful about how you deliver the material, people could end up confused, unable to recognize the value of futures, or completely aloof and disconnected by the end of the program. Being aware, flexible, iterative, and empathetic will allow you to design the right approach for your training program. I've taught some teams who were mandated (volunteered by their managers) to learn futures for their research and development work. While most of them were already working on future products and were using similar methods (even if they weren't aware that those methods were also part of the futures toolkit), some participants were not as engaged. I checked in constantly to make sure they were learning and finding the training useful and applicable to their work, but a combination of politics, skeptical attitudes, and distractions from their core jobs (pulling them away for several days or hours to be in a class was not favored by many who had important deadlines) made it difficult for me to be an effective mentor and facilitator.

AN ORGANIZATION IS CURIOUS ABOUT FUTURES AND IS ASKING YOU TO INTRODUCE THE TOPIC

In this situation, you can employ similar tactics to what you might do if you were working inside the organization. The client may already have an interest in or a notion of what futures is, but they may not know much about the process. So either you've done a great job selling it to them, or they've organically discovered it on their own. Either way, an invitation is a great way to start. Typically, I suggest introducing the topic with a one-hour presentation (a 45-minute lecture and 15 minutes of Q and A). This is a comfortable amount of time for anyone to attend, and not too much or too little time to make your point in. More than an hour could deter certain people from attending. You may even want to work with the client to make sure key leaders or influencers attend (in case you will need their financial or operational support later). If you fail to convince the client that they should invest in training or more consulting support from you, then you'll

---

4 "101" is commonly used in the US school system as the numerical designation for a basic introductory course in a subject or specialized field. A "101" course can include history, foundational principles, examples, or just a high-level overview of processes. A "102" or above course would be more advanced, exposing more complex systems, processes, and applications.

have the luxury of walking away and not having to live within their ecosystem with a failed attempt. You can always check back with them after a year or so to see if they are ready. Chances are, if people aren't convinced, it's usually because they can't relate the value to the immediate work they are doing today. They can't see how it impacts their bottom line, or they are still caught up in near-term tactics and metrics. That's why tailoring the content and language is so important to creating a common vocabulary that you can speak with. If you are successful, you may want to investigate why they want more so you can continue to target the aspects that do excite them.

Where and Why Futures Succeeds or Fails

As of the writing of this book, Futures Thinking is still a foreign concept to many. It still seems to operate on the fringes of design and strategy, but it is quickly gaining momentum in many sectors. There definitely is more evidence in the world to support using design thinking and service design methods to create successfully monetized and socially impacting products. And as these have become the exemplars for these approaches, they've become popularized as the pathways for how design-led thinking can lead to innovation and successful businesses. But even now, as IDEO (the founders and purveyors of Design Thinking) undergoes major changes (*https://oreil.ly/z7zme*) and human-centered design thinking is increasingly scrutinized for its anthropocentric, consumerist views of design, we are seeing new philosophies and approaches such as Futures Thinking and Life-Centered Design becoming adopted more widely and integrated into organizations outside of special units forced to think only about the far, far future. And this will probably be the case for a few more decades to come as some disciplines fade out (H1 horizons), and the H3 horizons merge into H2 horizons, and eventually H3 ideas become normalized (see "Bill Sharpe's Three Horizons" in Chapter 7). The great thing about watching all of this unfold is that there's still much to learn about the advent of new processes and how they may succeed or fail.

HOW FUTURES FAIL (COLLAPSE SCENARIOS)

I'll begin with the failures, because they are the barriers that are sometimes out of our control, and they either make us be more creative and diligent or detrimentally discourage us altogether. These are some of the indicators to look out for and nurture or avoid, if possible:

- Change or loss of leadership
- Focus on survival
- Lack of funding
- Competition with current strategic processes
- Refusal of other business units to adopt the plan
- Disintegration of excitement or change in interests
- Overly complex or dogmatic process or practitioners

Change or loss of leadership

Once the futures process is complete, a leader will be required to oversee the outcomes of the project. Those tasks could include managing a communication plan, implementing a strategic roadmap with clear near-term projects and tasks, building and sustaining the culture, or simply continuing to train, manage, and facilitate the process across the organization. If all goes well, everyone involved will follow and collaborate with the leader(s) to make futures work for the organization. They carry the knowledge, experience, documentation, network, resources, charisma, and negotiation tools to captain this potentially large and complex ship and oversee the many tasks that must continue for it to operate smoothly. If that leader leaves or is removed, you lose your captain and potentially all the knowledge, drive, and relationships they have as well. You could be dead in the water. Every team needs a captain to help manage and be the voice of inspiration and agent of change. We see this in many situations, such as when a CEO, vice president, or director leaves a team: the team undergoes a series of emotions that could eventually lead to attrition and loss of attention, excitement, and commitment to the work and the company.

Focus on survival

I consider startups to be some of the most difficult places to try and integrate futures. Startups, especially early-stage ones, may have a grand vision for where they want to go, but they will be spending most of their time trying to survive and prove they are a successful business model so they can get to the next stage of funding; they won't necessarily want to invest in an effort to map out a long-term strategy. However, this doesn't necessarily mean that it's an impossible mission. All product organizations have a strategy, be it short-term or long-term. In their world, "long-term" may equate to the next one to two years of survival. And depending on the product, they may need to address rapidly changing trends to survive. A survival mentality focused on immediate gains and learnings can be a tough one to work with. If you're committed, there are ways around it, but you'll need to be creative and resourceful and make a lot more analogies to near-term planning to sell futures work.

Lack of funding

Funding is always necessary. Whether it's your own personal project or one you want to do for a company, someone needs to pay for your time and to ensure the survival of the project. If you work for a company that barely has funding to keep itself alive, you may not have an opportunity to do anything that focuses beyond the next few months. There may not be enough money to invest in training, since all of the company's money is allocated to reaching the next quarter. But depending on the organization's appetite for a futures-enabled strategy, funding can always be found somewhere. The organization could pull in additional investors, take out loans, or defer payments until an agreed-upon future in which it could cover the bill.

Competition with current strategic processes

An organization might already have a strategic function, experts, and team dedicated to monitoring market trends or signals, but they might do it differently than we do. They might be happy with how that system works and may have garnered valuable successes from it, and so they don't want to rock the boat. They might even be territorial about the introduction of anything that sounds like "strategy." At most consulting firms, the business analysts are the strategists. Their degree is based on crunching numbers and creating business and market projections. If not undertaken cautiously, your campaign to introduce futures as a "strategic" process could fall on deaf ears—or worse yet, those who feel threatened could start a campaign to shut you down altogether. Without some kind of advocacy within those groups to open their minds to new tools and processes, you could be dead in the water from the start. Decide for yourself how much time and effort you're willing to put into that fight, and whether it's a fight worth fighting.

Refusal of other business units to adopt the plan

In the event that you need other departments, business units, or resources to help you execute the strategy (and you normally do need a lot of help to fulfill a long-term goal), you'll need to make sure they are all updated, committed, and excited to help. But if a team is not interested in helping, doesn't trust the process, and/or has no investment or "skin in the game" (*https://oreil.ly/DJ5Fd*), they may dismiss the vision or strategy for personal reasons or due to a lack of belief in the outcomes or requirements. Therefore, they could refuse to adopt the plan or to contribute the necessary resources or investments. In these cases, negotiation might be necessary, or you may need to find others who can provide what you need, or you may have to pivot the requirements as you reassess how to move forward without them.

Disintegration of excitement or change in interests

Sometimes, after the excitement fades away following the completion of a futures project, if the vision is not well communicated and internalized in such a way as to periodically remind people of it (whether through periodic updates, a giant poster of the vision on a wall, or planned meetings or events to reignite the excitement), commitment could slowly disintegrate, and people could forget about the vision and the strategic tactics that are being worked on to realize it. Sometimes people will stop and ask, "Why are we doing this again?" and you could face attrition or distractions. Sometimes the company's interests completely change. A new technology, trend, or climactic event emerges and dominates the media. Suddenly, even if you were able to see these weak signals in your research but no one took them seriously, the organization decides to shift all its resources toward something else. Eventually the strategy, its projects, and all of the funding and resources disappear, and a new strategy is rapidly developed (potentially without the use of a futures process). This happens often as organizations try to respond to market changes, especially if they see major threats to their business in the near term. Instead of you deeply analyzing the implications of threats and identifying logical and proactive opportunities to safely navigate the waters, the culture, leadership, and motivations can completely undermine everything you've worked hard to build. That's business, unfortunately. And this is why strong leadership, training, and an objective and open futures mindset are important for tackling these shifts as they arrive.

Overly complex or dogmatic processes or practitioners

This might be one of my biggest frustrations with working with experienced fore-sight and futures designers. People who have been educated in futures and have practiced it for many years naturally have a dedicated and sensitive inclination for *how* it should be practiced. There's nothing wrong with that. An artist may have a particular process that works for them, and they may have many customers that support them, so why change or advocate for anything else? However, to truly democratize futures, we need to be flexible and detach ourselves from overly strict dogma of how it's been practiced in the past. Saying you can't do it this way or that way simply creates a barrier against evolution of thought. Claiming territory for why something should be practiced one way or another, or forcing something complex on someone who can't totally grasp or understand it, will only end in distaste or an aversion to the craft. If I say that you could probably fix your car yourself and that I could show you how, but then I pull out a giant manual and say we must follow the directions, you might zone out completely and not even listen to half the things I'm saying, especially if it takes too long and is filled with jargon you don't understand. But if I just break it down into a few simple steps and use accessible language, you might be more willing to learn and still accomplish your goal. Yes, Futures Thinking can be complex at times. There are many factors to consider and so many tools you can use. And yes, there has been a lot of work done in the field trying to make sense of the future. But if your audience is not ready for it and you can't seem to boil things down into a language they can and want to understand, then you'll be shouting from your soapbox with little or no audience to hear you.

Adapting Your Program to Your Audience and Timeline

When training a pharma client recently, I was given only three hours a day over four days to complete an entire training program to teach them about Design Fiction. This was the term they had heard of and had been using to refer to what I called Futures Thinking and Speculative Design. All the participants were remote and from different countries (India, France, Japan). So I had to be careful not to use overly complex language or jargon, since English was not their first language. I negotiated with the client to start with a one-hour presentation as a primer on what Futures Thinking was, the stages and principles we use, and, using a variety of examples, what deliverables could look like. I showed them scenarios and

Futures Wheels culled from various projects, from academia to corporate software products, making sure to include examples that were closely related to their industry in healthcare and pharma so that they had something they could relate to as they imagined how the process might work for them. This gave the team a basis for what to expect and what the training program might look and feel like—the steps, the inputs and outputs. It was an accelerated journey through a selection of key principles, but I made sure to include plenty of time to practice with exercises as well as time to discuss, debate, and ask very specific questions about how they might apply Futures Thinking to their current culture, tools, processes, and projects. Scaling down the program, using many simple real-world analogies and an accessible vocabulary, and giving more room for discussion allowed us to tune the curriculum so that I could deliver an MVP (minimum viable product) of the futures toolkit in a way that felt practical and usable for them right away. Since these participants were not typically involved in the strategic planning process, I left out the stage about strategy but touched lightly on why it's important to have a strategy and connect today's actions toward future goals and visions. The three stages I used for this client were as follows:

1. I met with them to understand their business and immediate needs for Futures Thinking.
2. I delivered a presentation tailored to the audience and their business.
3. I designed a simplified program that traverses through the stages as applicable:
   a. How to find, analyze, and prioritize trends (research)
   b. How to use trends to develop scenarios and future worlds
   c. How to design innovative ideas in the future worlds

While this was more of a Futures 101 training program, it touched on the key stages they needed and was all I could really afford to do in the amount of time I was given. Though each day went by quickly, we had plenty of time to practice, which helped them learn by doing.

HOW FUTURES SUCCEEDS (TRANSFORMATIONAL SCENARIOS)

First and foremost, for futures to succeed, one has to be interested in exploring it. If your team, manager, client, or peers shut down the conversation too quickly, you may not have a chance, and it will be more work to convince them to open any doors for you. But if you encounter people with curiosity and an open mind, you're more likely to sway them. But what does *success* actually mean? You could certainly use some of the performance measurements I mentioned in Chapter 11. But the path to success can include the absence of or removal of barriers to make it work. And really, it takes leaders (top-down) and contributors (bottom-up) to see the value, and a desire to pursue it. It takes a collective effort, much like any project does. You can try to do things on your own, but designing the future should be a collaborative and inclusive affair. That said, there are some circumstances that are rich opportunity spaces for futures to work in whether or not it already exists within an organization. Here are just a few cases that can set you up for success:

- Strategic leadership that is invested in long-term thinking
- Teams that are interested in more strategic thinking
- Innovation and growth initiatives
- Long product cycles
- A product or service intent on large-scale change

Strategic leadership that is invested in long-term thinking

I say "strategic" leadership as opposed to just "leadership" to highlight the idea that a leader is actively involved in or cares about the strategy of a company. You might say, well, doesn't every leader have a strategy? Unfortunately, that's not the case. Some leaders are there to fix problems or just to keep the ship afloat. They may be trying to prove themselves because they're in the early stages of their career or role at the company, or they simply don't have an interest in thinking beyond survival mode. A strategic leader has a certain level of comfort, confidence, and excitement about planning for the future. They are always pushing the status quo and coming up with new ideas. They can see a signal or trend and internally extract it into the future to posit what it might mean for their organization or product. They might come up with too many ideas (some bad, some good), and that can either become really annoying or become real fodder for innovation. They also may be aware of the hardships and frictions of the current operation, but they are able to focus beyond what needs to be done today to succeed. When these kinds of leaders are interested and invested in new processes or ideas and are open to change and transformation, they can be your best friend. They'll clear the road for you to work. They will trust you as an expert and are more willing to invest in your support to facilitate innovation and long-term visioning.

Teams that are interested in more strategic thinking

Teams (business units, individual contributors, departments) that are already invested in long-term strategic thinking usually adopt futures faster. Whether it is product management or designers who know there is a clear need to think beyond near-term roadmaps, futures can be a way to open up the possibilities for them to expand their skill sets or begin influencing larger organizational or product strategies. Influence a team, and they can influence leaders to fund the effort. You may need to do a little bit of hand-holding to get them ramped up on the similarities and differences between how they are doing strategy today and what futures has to offer, but if you are speaking the same language already, then it should be a lot easier. Again, every culture and collective of personalities is different, and it really boils down to an excitement to be more creatively strategic. It isn't always just about process; sometimes it's about having that deep desire to try new things and wanting to make it work.

Innovation and growth initiatives

Innovation and growth initiatives can vary, but if there are any departments or efforts that are committed to nurturing innovation processes or pursuing moonshot projects (skunkworks (*https://oreil.ly/kAt_f*), research and development groups), these are groups you may want to partner with. Similarly, initiatives or groups that are assigned to developing the growth of the company are probably looking a few years further into the future and thinking strategically for the greater organization or teams. To grow anything takes time and naturally requires long-term thinking, so they are natural partners for the process. But "growth" can mean anything and can be tied to near-term metrics that don't necessarily incorporate long-term strategies. Try to look for the impacts and timelines being used on those projects and see if they can naturally align with your proposals. Any long-term initiative that intends to scale can be a little easier to convince.

Foresight at Disney Imagineering

In a discussion I had with Joe Tankersley, Strategic Foresight Lead at Disney Imagineering, he talked about some of the efforts and activities he was involved in during the early days of integrating foresight into Disney:

> In the Disney Human Resources group, and in the Imagineering group, foresight was seen as a skill that we wanted to spread throughout the organization, and particularly an imagineering process that made sense, because everybody who works in the creative part of Imagineering, they're all about imagining some future. So we were trying to get them to stretch the boundaries of what they would think about as being reasonable within the world that they lived in. At a place like Disney, they know what a theme park looks like. They know their theme parks work. And when you start to poke at the edges of that, the idea is to try to open up some new venues. And so one of the things we did was a lot of informal trend reporting—finding the information and distributing it, either through meetings with 30 or 40 people, or having two or three people share some trends and then have an open discussion about what this means.
>
> The biggest project we did in Imagineering was we spent about six months thinking about 2030. And we did a set of scenarios for 2030 that we thought were relevant, and we looked at the future of the city, the future collaboration, and themed entertainment. But that was definitely the least important in terms of what we did. We packaged up those scenarios and did a series of workshops where we would bring in 30 or 40 people to present the information and have them build specific artifacts and ideas which asked, what does this mean for us? What could we be doing? And a large question we asked was, if Imagineering decided they wanted to be in a different business in 2013, what would it be? Here's what the city looks like and here's what collaboration looks like. (This was around 2013, so we were looking 20 years into the future.)

It was interesting, because in that corporate environment, about 25–30% of the people just simply shut down when you started having this conversation: "It's too hard, I don't want to imagine it. My job is to get this widget from here to there. Not interested." And then you get about a third who are like, "I love this, I do this all the time. And by the way, we're going to change the world!" And then you get a big group of people in the middle who tend to walk away, and later they come back and say, "Hey, you know that thing you talked about? I saw that somewhere!" And for me, that was always the test of whether or not we were being successful. Could we at least get them to be more aware of what was going on outside of their world?

The big question was always, what's the value? People always want to know, "Well, this is great. This is interesting, but what does it have to do with what we do?" And so we actually spent about three months building a game that was designed to translate the idea of looking at trends into how you use trends to create strategy. It was a game/workshop kind of experience. And it was based on a theme park of the future. We played it with hundreds of cast members in multiple sessions. And we created a suite of scenarios in 10-year increments up to 2040. And for each of those increments, we would give you a set of new trends, but we put them in terms of "Here are the big trends, now you have to make some decisions. What do you want to spend your money on for the next 10 years if this is your theme park?" We talked about energy, equality, and all of those other domain trends you would expect to see. All of this was really designed to influence strategy. Once we went through the workshops, we went back and said, here's what your designers and creators are telling you. We actually had somebody who was part of our group who was also in the strategic planning group, who took it back and tried to say, hey, here are some of the things we're talking about, which I think is critical that you should also start thinking about.

People loved it. It was probably one of the most effective tools in terms of engaging people more playfully than some of the other kinds of methods. Disney was kind of serious internally but was open to that kind of serious play. After I left Disney, I became a consultant; I've worked with big and small corporations. And I will say that I think the one unique advantage of Disney is that they understand storytelling. So any time we presented anything, like the game, we felt like we were telling a story and it resonated with them, because it was part of their culture. I discovered later in other places that there was that interesting obstacle. But the minute you start talking about potential stories of the future, they become engaged, and it became a real advantage.

Long product cycles

Any industry that has products or services that require long product cycles—for instance, automotive, medical, industrial, and highly regulated industries take a long time to experiment, build, test, and deploy—is a great opportunity for Futures Thinking.[5] In the case of space exploration, it can take decades of speculation and prototyping on the ground to fully understand the implications of surviving the harsh conditions of space; to this day, we are still learning about the effects of long-term spaceflight on humans. Jet engines have around a 10-year product cycle. From concept to deployment, engineers can work on engineering ideas for several years before they finally see the light of day. These long cycles and timelines are already employing some form of long-term strategic planning, so they are a great place to start conversations about additional tools and ways to prototype the future. This isn't to say that startups or companies with short product cycles can't use futures (I mentioned this in Chapter 11 in the section on innovation accounting); they certainly can, but if you look for these longer-term cycles, they are great opportunities to introduce this process.

A product or service intent on large-scale change

Government or public service organizations naturally have longer timelines to implement their products or services. While there are certainly some digital products that can be deployed very quickly, there are still many efforts that require a lot of planning and policy development to become a reality. Agencies

---

5 This example is a general reference to organizations whose cycles require the use of robust long-term thinking systems and frameworks, but organizations with short cycles can absolutely use these same approaches to monitor, prepare, and strategize for near- or long-term futures.

that are responsible for services such as urban planning, public health and safety, agriculture, or education benefit from long-term planning in order to improve or innovate and keep up with society's needs. In 2014, for instance, the United Nations created the Sustainable Development Goals (SDGs), a collection of 17 interlinked objectives designed to serve as a "shared blueprint for peace and prosperity for people and the planet, now and into the future." The SDGs emphasize the interconnected environmental, social, and economic aspects of sustainable development by putting sustainability at their center (Figure 12-3). These goals seek to solve some of the hardest problems our species faces today. The sheer scale and effort it takes to inspire countries to work together toward long-term impact metrics makes them a natural fit for futures work.

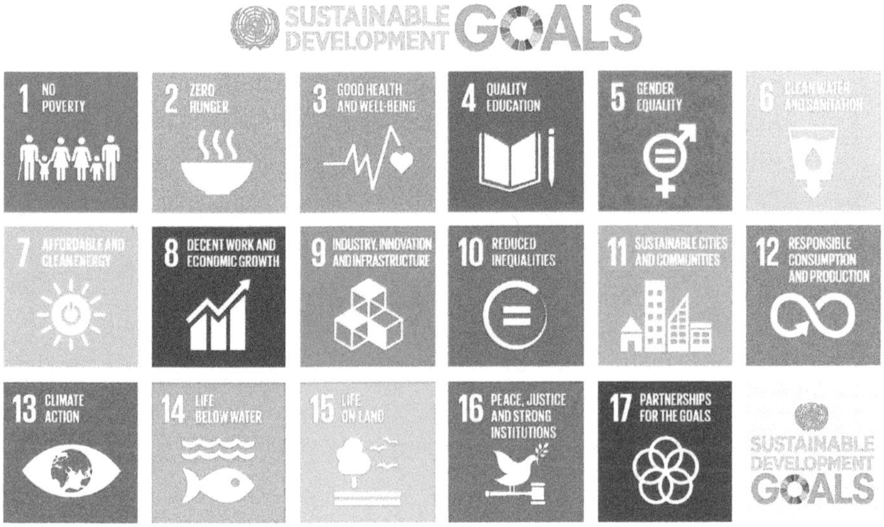

Figure 12-3. United Nations Sustainable Development Goals

The United Nations Department of Political and Peacebuilding Affairs (UN DPPA) Innovation Cell, headquartered in New York City and led by Martin Waehlisch, has fully embraced futures and foresight. As a peacekeeping organization, the UN must anticipate conflict, natural disasters, and any event that could disrupt peace and security among or within nations. It has several efforts to train its staff and global ambassadors to think creatively about how to use technology in field operations as well as within peacebuilding negotiations. But the road to integrating futures within the greater UN has not been an easy one, even though part of its job is to anticipate and assist in developing a lasting and peaceful future for its member states.

Innovation and Foresight at the UN DPPA

During an interview with Martin Waehlisch, founding member of the UN DPPA Innovation Cell, and Minji Song, political affairs officer in the UN's Department of Political and Peacebuilding Affairs, Waehlisch explained some of the successes and challenges of building foresight capabilities at the UN DPPA:

> When working on introducing futures into the UN, we were trying to find the right semantics for Futures Thinking. Minji Song, a UN political affairs officer, says she's a **futures enthusiast**. So we have foresight experts and futures enthusiasts. The UN uses the term **Strategic Foresight**, which in the UN context is basically dominated by military patents. And we have a lot of military folks who do planning. But because of this confusion in semantics, our leadership has almost ridiculed Futures Thinking. They've said, "Well, this is what we do anyway—we make sense of the world. Full stop. Why do you call this Strategic Foresight or foresight in small letters or capital letters? That's confusing, you're just watering down something we already do! And why do you give it a special name?" Thus the confusion around the terminology has made it difficult for us to make space for Futures Thinking in some parts of the organization.
>
> It demonstrates the complexity and the multiple lenses that you can apply to this complexity. We also have lots of experience with **pop futurists**; we call them pop futurists because they are these very visible consultants and figures who are talking about foresight methodologies, as if this is a very difficult scientific discipline. But for the UN context, at least how our senior management was thinking, we wanted it to be more applicable to our needs. And I think that's really because there hasn't been a very clear and documented impact from a foresight process. But of course conflict has been prevented because of a Strategic Foresight process where it was clear that it was because of that

exercise. And of course, it's never because of one factor that conflict prevention happens. At the global level in the UN, we are having lots of conversations about the whole system's transformation. And this is called the **quintet of change***. And this is a priority strategy for the Secretary General. They want the whole human rights organization to change its culture, its capabilities, be more inclusive, and one of the five pillars is the use of strategic foresight. But we had to negotiate the title. Would we refer to it as Strategic Foresight or just foresight (lowercase)?*

There's a lot of space to think about and apply foresight in managing technological disruptions in our work in peace and security. And I think the Innovation Cell is very good at putting this on the agenda. So something like a simple trends report and translating that in a manner that senior managers can understand it in terms of the impact for business security, these are very important, but again, the apparatus that we are given to deal with the disruption becomes incredibly political—it becomes a political conversation about who deals with it. How do the member states deal with it? Are we positive that there will be consequences to the peace and security of the world? Is it just peace and security or is it the whole of humanity? Should the UN DPPA be looking at it? And those kinds of considerations about thinking about who owns this thinking and propositions, the bureaucracy and where it should be placed, means that we're five years late already.

Future thinking needs leadership. Because, the future. It's something you do. Intuitively, it's part of your mindset, your literacy. It requires brave individuals that are not shy to test new lessons, experiment, and actually make futures matter. I think we can be proud of cases where we provoked people to think about things upside down. We generated provocative scenarios, not just the usual best case/worst case scenario, and that was learning by doing, as futures is always learning by doing. And these were able to "stress test" some of the planning assumptions the UN had in certain regions for various peace efforts.

We had a recent youth project in Northeast Asia where we were using futures thinking and speculative design as a way to engage youth peacebuilders on very complex issues related to the security architecture and the political future of that region. Typically, when you put people around a table and ask them, "What's the future like?" especially in highly political settings, like the United Nations, or in the setting of International Affairs, either people just start reading from their own script, or they resort back to old thinking and are very conservative. We were faced with the challenge of how do we get people out of their comfort zone and bring them into a space where they can feel like they're allowed to be creative? So network training and guidance was useful to create a kind of container for them to feel like they have the skills, the language to have those conversations and for them to feel confident to also add those kinds of visions and to challenge other people's visions. That container was very important. And we've also provided the container with a kind of full ownership so that they felt that they were encouraged to discuss and really push them forward. And this was the case with some of the speculative designers that we worked with who had that training from speculative design experts. And you also had that container through the kind of the networks provided by the academic institution to freely be creative.

In Closing

Every organization has its cultural nuances that you'll need to consider when trying to integrate Futures Thinking into their work or daily lives. Think about where you want to start, the goals and outcomes you want, and how you want to lead and facilitate that roadmap. You can treat each effort as a futures strategy in itself, collecting evidence of history (what worked in the past), patterns that are driving innovation or transformation today (what is moving the company forward), and aspirations (where the company wants to go). You can map out implications and scenarios based on different tactics you want to try and think about what will happen if you succeed or fail at any of them. All the while, you can have a long-term vision for how to integrate the process and begin bringing measurable value to your team, your stakeholders, your investors, or your organization's strategy. In the end, it will be a journey and adventure worth trying. If you really believe in the value of Futures Thinking, then give it a shot; if you fail, you can determine whether you need to go somewhere else, try something different, or just drop the effort altogether. It's up to you to decide whether it will be acceptable to carry on a mission if the hurdles are too high or the reward is too little or there's yet another mountain to climb. But the reward of successfully making futures work for yourself and others is priceless. As ambassadors and facilitators of the future, we all have a special gift and responsibility to society and our planet; with futures, we can transform how we live, work, and survive while manifesting our greatest dreams and desires. So use it wisely. And let's make futures work.

Afterword

This afterword is written by Ben Lowdon, a practicing design futurist, design director, and coach at eBay, having worked previously at Zalando and Nokia (future maps). Ben is concerned with how design helps to create better outcomes for people and the planet and uses futuring methods within his coaching and professional design leadership practice.

Next Steps: Continuing Your Journey

We're at the end of this futures primer, and you have now acquired a new mindset and a big bag of new tricks. You've learned how design plays a crucial role in futures and foresight and how it can assist in discovering, synthesizing, and facilitating information into alternative visions, provocations, and solutions. You've also learned how futures thinking expands the range of typical design research methodology and practice and how you might apply this mindset to the entirety of your process. You've discovered new tools to practice and learned about the importance of developing the right outlook and practicing imagination and curiosity for the future. But what now?

Well, having been through this step myself, I can confidently tell you that you just need to begin by actively applying what you have learned here. You could start with picking a personal project and working on it. Maybe find a few people at your workplace whom you can safely practice and try different tools with and get feedback on your approach. Start or join a broader community of practice of people you can share your passion and ideas with. There are many futures communities and resources out there already that you could tap into. Join one of these inspirational communities and fully immerse yourself. This immersion and practice will help you to deepen your knowledge and importantly understand which parts of the futures toolkit best suit your needs and practice.

Apply Futures Thinking to Your Own Life

Another way to begin experimenting is to start looking at your own future. I find it a good rule that many of the strategies you apply within your work can also be applied to your own life and career strategy. So let's return to our sailing analogy and consider: If you think of your life as a sailboat, then where are you headed? Where is the wind? What are the rocks and obstacles in your way? And do you (even) have a map?

The mindset and tools shared here can help us all understand and nudge our lives toward futures we might personally want to live in, or even create. You can now look at the uncertainty (VUCA) within your life and better assess where you need to pay attention, what to mitigate, and where you need to act. You can also use the mindset to speculate and actually prototype (hello, designers!) the different futures you or your family might like to live in. Dreaming, and being strategic and (act)ively invested in creating your own future, is not only self-affirming but also a great space to explore the tools of futures practice while building quality and resilience into your life.

For me, this is a longer lens that I (belatedly) apply to my life, family, and career, and also when coaching and managing my teams. I now take time to identify potential partners and cultivate conversations with people to imagine where we might want to be in x years and how that knowledge might inform the decisions that they make today. I try to help them zoom out and see the narrative of their lives, and the potential implications and quality of outcomes of our decisions and efforts. Someone who knows where they are and who has a clear mission is much more engaged, and also—let's face it—interesting to work, create, or even live with. So where do you want to be in 10 years? How might you look back on your decisions today in 50 years and know that you had been a better ancestor—one who had built generational knowledge, wealth, or equity? The reality is that this is a powerful toolkit to enrich and influence outcomes, and it's valuable for your life and career too.

Complementary Approaches

Now let's zoom out and consider how you might use futures to create impact for your social group or industry or even the planet.

Futuring is a deep discipline that you can of course specialize in (and even become very dogmatic about), but more likely it will become a complement to the many other approaches that you've been learning in your design practice. If you look broadly at the world, you can see examples of this already—futurists and designers whose specialism or training might be elsewhere, but who use futures and foresight to improve the outcomes of the topics that they are committed to on a systems level, which might include health, food, sustainability, politics, or any of numerous other examples.

I suggest you examine different design futurists' work and study their processes to understand how they create impact. There are of course many good examples, but a good start is to look at designers like Monika Bielskyte (protopias), Leyla Acaroglu (circular futures), Damien Lutz (planet-centered design), Cennydd Bowles (future ethics), and Daniel Wahl (regenerative design). These are all people who will open the door to other, deeper rabbit holes. But I would go further and suggest that these designers have become actively engaged and invested in their futures. And while working with Phil over the years, I have observed that, while being an active designer, he is also engaged and invested in the future of design, and in helping designers be more strategic and curious about the things they put into the world.

Another good example of complementary disciplines is Transition Design. Transition Design is a practice created by Terry Irwin, Cameron Tonkinwise, and Gideon Kossoff at Carnegie Mellon and is concerned with how we catalyze societal transitions toward more sustainable and inclusive futures (Figure A-1). In short, it is focused on how we work to give people and the planet equity and a healthy, more secure future. To radically oversimplify, Transition Design is a mix of expertise that you might already have acquired, such as: (1) Design Thinking and (2) Systems Thinking, enriched with (3) sustainability expertise and (4) change management, and finally (5) Futures Thinking or foresight.

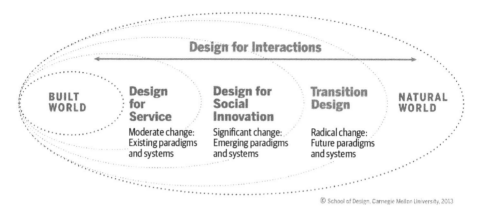

Figure A-1. Transition design in context (source: Carnegie Melon University, 2013)

In this instance, futuring is what helps to articulate and understand the outcomes of different scenarios and engage participants in imagining radically transformational futures that respect human and planetary boundaries. These five skills together form a super toolkit to affect change on a local and planetary scale and are a great example of how futures should be combined with other skill sets for greater impact. Like many big shifts in design practice, this new way of thinking starts in academia and shifts into the industry. I personally hope that our next generation of designers will assume that this way of thinking is just the norm.

New Responsibilities

Ultimately, Design Futures is about imagining more desirable futures (what desirable means to you is a strictly personal question). And for us as designers, that means understanding and mitigating the impact of the objects, imaginaries, and possibilities that we introduce into the world. From this point on, everything you design has a resonance and a potential to create bigger ripples. You now have the tools to understand the intentions of the things that you put into the world and identify whose future you are improving or degrading, and also consider who or what is left out of your picture. In many ways I hope that this is an event horizon in your career, as from this point on everything that you design and share must be considered intentional. Everything is now a decision, a provocation, and an opportunity. You get to raise the bar and collectively build new narratives and visions of the future. Consider: what might it look like if you and your peers were to design the next postcards of a 100-year future? Which of those new imaginaries might we want to put into the world?

To wrap up—there are enough people proclaiming about the future of designers, so I would rather share an expectation. I expect that as a successful "strategic designer" in 2034, you will have a largely different toolkit than today. You will understand how to scan an emergent and changing future and will help your team or community to manage constantly emerging complexity. You will know how to build equitably and sustainably (maybe even regeneratively) and actively do less harm. But I also expect that like service thinking, or user research, futures (and systems) thinking will have become an integrated part of every designer and facilitators toolkit. If we're all successful, you might not even differentiate it from your usual practice. That you're reading this now means that you are taking the right steps and are on your way to creating more intentional, inclusive, and resilient futures and being a better designer. Practice, share your work, and give this book to someone else to read once you're finished with it.

and your peers were to design the next postcards of a 100-year future? Which of those new imaginaries might we want to put into the world?

To wrap up—there are enough people proclaiming about the future of designers, so I would rather share an expectation. I expect that as a successful "strategic designer" in 2034, you will have a largely different toolkit than today. You will understand how to scan an emergent and changing future and will help your team or community to manage constantly emerging complexity. You will know how to build equitably and sustainably (maybe even regeneratively) and actively do less harm. But I also expect that like service thinking, or user research, futures (and systems) thinking will have become an integrated part of every designer and facilitators toolkit. If we're all successful, you might not even differentiate it from your usual practice. That you're reading this now means that you are taking the right steps and are on your way to creating more intentional, inclusive, and resilient futures and being a better designer. Practice, share your work, and give this book to someone else to read once you're finished with it.

Glossary

Adversarial Design (https://oreil.ly/XADlg)
> Adversarial Design is a type of political design that evokes and engages political issues. In doing so, the cultural production of Adversarial Design crosses all disciplinary boundaries in the construction of objects, interfaces, networks, spaces, and events. Most importantly, Adversarial Design does the work in expressing and enabling agonism. (See also: Carl DiSalvo, *Adversarial Design* [Cambridge, MA: MIT Press, 2012].)

Applied Futures (https://oreil.ly/EVkoe)
> Applied Futures Thinking is a construct of several theories and approaches that allow us to help organizations in making future perspectives actionable. Most notably, applied futures thinking is a form of futures studies: the study of possible, probable, and preferable futures.

augmented reality (https://oreil.ly/Idtbo)
> Augmented reality (AR) is the real-time use of information in the form of text, graphics, audio, and other virtual enhancements integrated with real-world objects. It is this "real world" element that differentiates AR from virtual reality. AR integrates and adds value to the user's interaction with the real world, versus a simulation.

backcasting (https://oreil.ly/XN2WW)
> Backcasting is a planning method that starts with defining a desirable future and then works backwards to identify policies and programs that will connect that specified future to the present. The fundamentals of the method were outlined by John B. Robinson from the University of Waterloo in 1990. The fundamental question of backcasting asks: "If we want to attain a certain goal, what actions must be taken to get there?"

black swan

> *Black swan* is a term used to describe a highly improbable and unpredictable event that has a major impact. The concept was popularized by Nassim Nicholas Taleb in his book *The Black Swan* (Random House). Black swan events are characterized by their rarity, their extreme impact, and the tendency of people to rationalize them after the fact. Examples of black swans include financial meltdowns, pandemics, natural disasters, technological disruptions, and terrorist attacks.[1]

Causal Layered Analysis (https://oreil.ly/FZjot)

> Causal Layered Analysis (CLA) is a future research theory that integrates various epistemic modes, creates spaces for alternative futures, and consists of four layers: Litany, Social and Structural, Worldview, and Myth/ Metaphor. The method was created by Sohail Inayatullah, a Pakistani-Australian futures studies researcher.

Critical Design (https://oreil.ly/7cXgz)

> A term first used by Anthony Dunne in his book *Hertzian Tales* (Royal College of Art, 1999), referring to an attitude toward design rather than a movement or method. It follows in the footsteps of other practices (like Radical Design in Italy and avant-garde British architecture of the late 1960s and early 1970s) that have regarded design as a way to pose incisive questions, challenge the status quo, and think deeply about the possible future consequences of present choices. Critical Design is speculative, conceptual, provocative, and can be darkly satirical. It does not always lead to usable products, but it does produce long-term thinking, a nuanced view of consumers as complex, contradictory individuals, and alternative solutions suggesting that change is always possible, even inevitable.

Delphi method

> Forecasting technique that consists of an exercise in group communication among a panel of geographically dispersed experts. It facilitates the formation of a group judgment without permitting the type of social interactive behavior that occurs during a normal group discussion. It is essentially a method for achieving a structured anonymous interaction between carefully selected experts by means of a questionnaire with controlled feedback.

---

1 Description written with the assistance of ChatGPT.

Design Fiction (https://oreil.ly/zce1m)

Design Fiction is a design practice aiming at exploring and criticizing possible futures by creating speculative, and often provocative, scenarios narrated through designed artifacts.

Design for Debate (https://oreil.ly/qeR5w)

Design today is concerned primarily with commercial and marketing activities, but it could operate on a more intellectual level. It could place new technological developments within imaginary but believable everyday situations that would allow us to debate the implications of different technological futures before they happen. This shift from thinking about applications to implications creates a need for new design roles, contexts, and methods. It's not only about designing for commercial, market-led contexts, but also for broader societal ones. It's not only about designing products that can be consumed and used today, but also imaginary ones that might exist in years to come. And it's not only about imagining things we desire, but also undesirable things—cautionary tales that highlight what might happen if we carelessly introduce new technologies into society.

design probes

In her 2018 bachelor's thesis "Design Probes: A Good Method for Designing with Children" *(https://oreil.ly/gdBof)*, Sonja Rönnberg defined design probes as a user-centered design method that "enable[s] dialogues with users, focusing on giving the participating users inspiration. Inspiration is considered to be more valuable than information and is the designer's responsibility to carry through in the creation of design probes. It is also up to the designer to interpret the users' world in a fashion that does not validate or evaluate it."

diegetic prototype

The "diegetic prototype," according to David Kirby, is a design or object that exists in the fictional world and is created through "dialogue, plot rationalizations, character interactions and narrative structure."[2]

---

2 David Kirby, "The Future Is Now: Diegetic Prototypes and the Role of Popular Films in Generating Real-World Technological Development" *(https://oreil.ly/6KkYn)*, *Social Studies of Science* 50, no. 1 (2010): 41–70.

Discursive Design

Discursive Design is a type of design that operates or is sometimes synonymous with other forms of anticipatory or speculative design, such as Critical Design and Design Fiction. In a 2015 article about Discursive Design (*https://oreil.ly/tD9fz*), the designers Bruce and Stephanie Tharp expressed their belief that the term *discursive* "represents the core of all of these forms, and operates most effectively as an organizing genus. The word 'discursive' comes from 'discourse,' which can be understood most basically as the expression or treatment of a topic—part of a discussion or debate. A discursive design is an object that has been intentionally (and usually abstractly) embedded with discourse and/or is used to elicit discussion."

Double Diamond (https://oreil.ly/FoK6i)

Double Diamond is the name of a design process model popularized by the British Design Council in 2005 and adapted from the divergence-convergence model proposed in 1996 by Hungarian-American linguist Béla H. Bánáthy. The two diamonds represent a process of exploring an issue more widely or deeply (divergent thinking) and then taking focused action (convergent thinking). (For more on the four phases of the Double Diamond—Discover, Define, Develop, Deliver—see the Design Council website (*https://oreil.ly/etTKt*).)

Experiential Futures Ladder

Devised by Stuart Candy and Jake Dunagan, the Experiential Futures Ladder is a framework for immersive experience that is designed to situate its audience into future situations or contexts through designed physical (or digital) experiences. In a post on his blog *The sceptical futuryst* (*https://oreil.ly/WEOoB*), Candy observed that "[m]ost traditional futures practice, and certainly scholarship, operates on a high level of abstraction, above the experiential threshold, while experiential work explores more concrete manifestations of futures—possible, probable and preferable."

fad

A fad is the shortest-lasting type of trend. Fads fade as quickly as they appear. Traditionally they have been associated with fashion fads, but in recent years they are correlated with viral trends that appear on social media platforms.

focal issue (focal topic) (https://oreil.ly/lSUC3)

A focal issue is the focus of a foresight initiative—it largely depends on the objective of the project, as well as the culture of the community or organization that the initiative aims to impact. Some examples of common focal issues include "The Future of Education," "The Future of Retail," or "The Future of AI."

forecasting

Forecasting is making a prediction based on events in the present and past, in addition to the observation of trends.

Futures Cone

In a 2017 post on his blog *The Voroscope* (*https://oreil.ly/smt_S*), Joseph Voros explained that "[t]he 'futures cone' model was used to portray alternative futures by [Trevor] Hancock and [Clement] Bezold (1994), and was itself based on a taxonomy of futures by [Norman] Henchey (1978), wherein four main classes of future were discussed (possible, plausible, probable, preferable)." Voros expanded the cone to describe seven types of alternative futures: Potential, Preposterous, Possible, Plausible, Probable, Preferred, and Projected. (For more discussion of Voros's range of alternative futures, see Epaminondas Christophilopoulos, "Special Relativity Theory Expands the Futures Cone's Conceptualisation of the Futures and the Past" (*https://oreil.ly/6SEab*), *Journal of Futures Studies* 26, no. 1 [September 2021]: 83–90.)

futures persona (https://oreil.ly/XCOhn)

Futures personas or future personas are fictional individuals living in future scenarios. They are created from scenarios. With their cognition and behavior, they embody the scenarios. They are the living essence of the futures.

Futures Studies (https://oreil.ly/UJeAN)

Futures Studies, or futures research, is the systematic study of possible, probable, and preferable futures. The field has broadened into an exploration of alternative futures and deepened to investigate the worldviews and mythologies that underlie our collective prospects.

Futures Triangle (https://oreil.ly/yTnrE)

The Futures Triangle is a tool developed by Dr. Sohail Inayatullah in 2008 that not-for-profit CEOs and boards can use to create a frame from within which to view the current forces at work on the organization's planning and decision making for impact and future growth. It's a great place to start a foresight project as it can bring to life the different forces at play.

Futures Wheel (aka Implications Wheel) (https://oreil.ly/7UbM1)

The Futures Wheel is a method for graphical visualization of direct and indirect future consequences of a particular change or development. It was invented by Jerome C. Glenn in 1971.

indicator (https://oreil.ly/v4Ut-)

Indicators are tools used to measure and assess progress or performance in a particular area. In monitoring and evaluation, there are several types of indicators that can be used to measure progress and success: input indicators, output indicators, outcome indicators, process indicators, impact indicators, efficiency indicators, performance indicators, context indicators, and sustainability indicators.

Interrogative Design (https://oreil.ly/OSo-8)

Most commonly associated with the work of artist Krzysztof Wodiczko. His practice, known as Interrogative Design, combines art and technology as a critical design practice in order to highlight marginal social communities and add legitimacy to cultural issues that are often given little design attention.

Ludic Design

The website for Coventry University's Centre for Postdigital Cultures (*https://oreil.ly/Ifcn7*) explains that "Ludic Design focuses on the transdisciplinary investigation into playful and gameful design, practice, and culture with impact on socio-cultural development. It includes studies into the mechanics, dynamics, and aesthetics of play and gameplay for informing meaningful experience design. The wide reach of play and gameplay is universal, providing opportunities for ludic design aspects to be embedded into mediums for communicating serious messages, affecting emotions in engaging stories and environments, and impacting attitudes and behaviours in the process."

focal issue (focal topic) (https://oreil.ly/lSUC3)

A focal issue is the focus of a foresight initiative—it largely depends on the objective of the project, as well as the culture of the community or organization that the initiative aims to impact. Some examples of common focal issues include "The Future of Education," "The Future of Retail," or "The Future of AI."

forecasting

Forecasting is making a prediction based on events in the present and past, in addition to the observation of trends.

Futures Cone

In a 2017 post on his blog *The Voroscope* (*https://oreil.ly/smt_S*), Joseph Voros explained that "[t]he 'futures cone' model was used to portray alternative futures by [Trevor] Hancock and [Clement] Bezold (1994), and was itself based on a taxonomy of futures by [Norman] Henchey (1978), wherein four main classes of future were discussed (possible, plausible, probable, preferable)." Voros expanded the cone to describe seven types of alternative futures: Potential, Preposterous, Possible, Plausible, Probable, Preferred, and Projected. (For more discussion of Voros's range of alternative futures, see Epaminondas Christophilopoulos, "Special Relativity Theory Expands the Futures Cone's Conceptualisation of the Futures and the Past" (*https://oreil.ly/6SEab*), *Journal of Futures Studies* 26, no. 1 [September 2021]: 83–90.)

futures persona (https://oreil.ly/XCOhn)

Futures personas or future personas are fictional individuals living in future scenarios. They are created from scenarios. With their cognition and behavior, they embody the scenarios. They are the living essence of the futures.

Futures Studies (https://oreil.ly/UJeAN)

Futures Studies, or futures research, is the systematic study of possible, probable, and preferable futures. The field has broadened into an exploration of alternative futures and deepened to investigate the worldviews and mythologies that underlie our collective prospects.

Futures Triangle (https://oreil.ly/yTnrE)
The Futures Triangle is a tool developed by Dr. Sohail Inayatullah in 2008 that not-for-profit CEOs and boards can use to create a frame from within which to view the current forces at work on the organization's planning and decision making for impact and future growth. It's a great place to start a foresight project as it can bring to life the different forces at play.

Futures Wheel (aka Implications Wheel) (https://oreil.ly/7UbM1)
The Futures Wheel is a method for graphical visualization of direct and indirect future consequences of a particular change or development. It was invented by Jerome C. Glenn in 1971.

indicator (https://oreil.ly/v4Ut-)
Indicators are tools used to measure and assess progress or performance in a particular area. In monitoring and evaluation, there are several types of indicators that can be used to measure progress and success: input indicators, output indicators, outcome indicators, process indicators, impact indicators, efficiency indicators, performance indicators, context indicators, and sustainability indicators.

Interrogative Design (https://oreil.ly/OSo-8)
Most commonly associated with the work of artist Krzysztof Wodiczko. His practice, known as Interrogative Design, combines art and technology as a critical design practice in order to highlight marginal social communities and add legitimacy to cultural issues that are often given little design attention.

Ludic Design
The website for Coventry University's Centre for Postdigital Cultures (*https://oreil.ly/Ifcn7*) explains that "Ludic Design focuses on the transdisciplinary investigation into playful and gameful design, practice, and culture with impact on socio-cultural development. It includes studies into the mechanics, dynamics, and aesthetics of play and gameplay for informing meaningful experience design. The wide reach of play and gameplay is universal, providing opportunities for ludic design aspects to be embedded into mediums for communicating serious messages, affecting emotions in engaging stories and environments, and impacting attitudes and behaviours in the process."

macro trend (https://oreil.ly/11Z5L)

Macro trends refer to major shifts in consumer behavior that will direct the business landscape in the long term. They have a cross-industry impact and evolve over time. Examples of previous global macro trends include the adoption of social media and catering to the aging population.

Mad Libs (https://oreil.ly/vwDPl)

Mad Libs is a phrasal template word game created by Leonard Stern and Roger Price. It consists of one player prompting others for a list of words to substitute for blanks in a story before reading aloud. The game is frequently played as a party game or as a pastime. The game was invented in the United States, and more than 110 million copies of Mad Libs books have been sold since the series was first published in 1958.

mega trend (https://oreil.ly/gYfRd)

Mega trends are the great forces in societal development that will very likely affect the future in all areas over the next 10–15 years. A mega trend is also defined as a large, social, economic, political, environmental, or technological change that is slow to form. Once in place, mega trends influence a wide range of activities, processes, and perceptions, both in government and in society, possibly for decades. They are the underlying forces that drive trends (i.e., aging population).

metaverse (https://oreil.ly/2rO3c)

The metaverse is a loosely defined term referring to virtual worlds in which users represented by avatars interact, usually in 3D and usually focused on social and economic connection. The term *metaverse* originated in the 1992 science fiction novel *Snow Crash* as a portmanteau of "meta" and "universe." In *Snow Crash*, the metaverse is envisioned as a hypothetical iteration of the internet as a single, universal, and immersive virtual world that is facilitated by the use of virtual reality (VR) and augmented reality (AR) headsets.

micro trend (https://oreil.ly/11Z5L)

Micro trends refer to smaller/niche shifts in consumer behavior that are changing the business landscape in the short term. They are usually linked to the underlying values and behaviors associated with macro trends.

mixed reality (https://oreil.ly/CabZP)

Mixed reality (MR) is a term used to describe the merging of a real-world environment and a computer-generated one. Physical and virtual objects may coexist in mixed reality environments and interact in real time.

NFT (nonfungible token) (https://oreil.ly/Yf--m)

A nonfungible token (NFT) is a unique digital identifier that is recorded on a blockchain and is used to certify ownership and authenticity. It cannot be copied, substituted, or subdivided. The ownership of an NFT is recorded in the blockchain and can be transferred by the owner, allowing NFTs to be sold and traded.

Probability Versus Impact Matrix (https://oreil.ly/EcmQW)

A tool used in qualitative risk analysis to evaluate the likelihood and impact of identified risks. It uses a numerical scale to rate the likelihood and impact of each risk, which can then be used to prioritize risks and develop strategies to mitigate or manage them.

prompt cards

Prompt cards are used to provoke, inspire, and stimulate conversation. They generally can take the form of audio, visual, haptic, or word prompts and can represent any category of subject or topic. Prompt cards or games are typically used in ideation and brainstorming activities to stimulate creativity.

prototype (https://oreil.ly/DwkY_)

A prototype is an early sample, model, or release of a product built to test a concept or process. It is a term used in a variety of contexts, including semantics, design, electronics, and software programming. A prototype is generally used to evaluate a new design to enhance precision by system analysts and users. Prototyping serves to provide specifications for a real, working system rather than a theoretical one. In some design workflow models, creating a prototype (a process sometimes called *materialization*) is the step between the formalization and the evaluation of an idea.

provocation

A provocation is defined as "something that incites or provokes; a means of arousing or stirring to action" *(https://oreil.ly/Q3flX)*. In terms of Futures Thinking, provocation is used to describe future concepts or ideas that are meant to stir up debate, spark conversation, or incite change.

scenario planning (https://oreil.ly/7gXap)

> *Scenario planning, scenario thinking, scenario analysis, scenario prediction,* and the *scenario method* all describe a strategic planning method that some organizations use to make flexible long-term plans. It is in large part an adaptation and generalization of classic methods used by military intelligence. In the most common application of the method, analysts generate simulation games for policymakers. The method combines known facts, such as demographics, geography, and mineral reserves, with military, political, and industrial information, and key driving forces identified by considering social, technical, economic, environmental, and political ("STEEP") trends.

Science Fiction Prototyping (Sci-Fi Prototyping) (https://oreil.ly/gCKsh)

> Science Fiction Prototyping (SFP) refers to the idea of using science fiction to describe and explore the implications of futuristic technologies and the social structures enabled by them. The idea was introduced in 2010 by Brian David Johnson, who, at the time, was a futurist at Intel working on the challenge his company faced anticipating the market needs for integrated circuits at the end of their 7- to 10-year design and production cycle.

signal

> A signal is typically a small or local innovation or disruption that has the potential to grow in scale and geographic distribution. A signal can be a new product, a new practice, a new market strategy, a new policy, or new technology. It can be an event, a local trend, or an organization. It can also be a recently revealed problem or state of affairs. In short, it is something that catches our attention at one scale and in one locale and points to larger implications for other locales or even globally.

Speculative Design (https://oreil.ly/Mp4l9)

> Speculative Design is a methodology for training one's ability to look to the future and explore complex problems. Working with Speculative Design helps us explore new perspectives and gives us tools and support to use our imagination to both imagine and shape possible futures.

stakeholder map (https://oreil.ly/25c2q)

A stakeholder map is a visual or physical representation of the various individuals and groups involved with a particular challenge or system. Stakeholders are people, groups, or individuals who have the power either to affect or be affected by the design project you're involved in. They range from the head of your organization to your core group of users, to the man in the street who may just experience a few effects of what you set out to do. Your stakeholders also include people you communicate with throughout the project, report insights to, and carry out activities with.

Strategic Foresight

Strategic Foresight, also referred to simply as foresight, is the action of predicting what will happen or will be needed in the future.

subject matter expert (https://oreil.ly/G_U-J)

A subject matter expert (SME) is a person who has accumulated great knowledge in a particular field or topic and this level of knowledge is demonstrated by the person's degree, licensure, or years of professional experience with the subject.

time horizon (https://oreil.ly/sZSaq)

A time horizon, also known as a planning horizon, is a fixed point of time in the future at which point certain processes will be evaluated or assumed to end.

trend

A trend is a pattern we observe, or the general tendency or direction of a development or change over time.

trend drivers (https://oreil.ly/zZUAB)

Trend drivers (aka drivers) are the series of underlying causes and forces that shift new developments in a particular direction. They are what influence and trigger the change. By studying the trend drivers in multiple local and global contexts, a trend forecaster can consider how they will influence lifestyles and ultimately the offering of products and services in the marketplace. Trend drivers are not mutually exclusive; all are interwoven with complex cause-and-effect relationships.

virtual reality (https://oreil.ly/vsWhQ)

Virtual reality (VR) is a simulated experience that employs pose tracking and 3D near-eye displays to give the user an immersive feel of a virtual world. Applications of virtual reality include entertainment (particularly video games), education (such as medical, safety, or military training), and business (such as virtual meetings). VR is one of the key technologies in the reality-virtuality continuum. As such, it is different from other digital visualization solutions, such as augmented virtuality and augmented reality.

weak signal (https://oreil.ly/qe3Xn)

A weak signal is the first indicator of a change or an emerging issue that may become significant in the future. Weak signals are often identified as a part of horizon scanning (environmental scanning) that supplements trend analysis and can be used as a foundation for defining wild cards.

Web 3.0

Web 3.0 refers to the next generation of the World Wide Web that is characterized by advanced technologies and a more intelligent, interconnected, and decentralized internet. While there is no universally agreed-upon definition, Web 3.0 is often envisioned as a set of emerging technologies and paradigms that aim to enhance the way information is accessed, shared, and processed on the internet.

virtual reality (https://oreil.ly/vsWhQ)

Virtual reality (VR) is a simulated experience that employs pose tracking and 3D near-eye displays to give the user an immersive feel of a virtual world. Applications of virtual reality include entertainment (particularly video games), education (such as medical, safety, or military training), and business (such as virtual meetings). VR is one of the key technologies in the reality-virtuality continuum. As such, it is different from other digital visualization solutions, such as augmented virtuality and augmented reality.

weak signal (https://oreil.ly/qe3Xn)

A weak signal is the first indicator of a change or an emerging issue that may become significant in the future. Weak signals are often identified as a part of horizon scanning (environmental scanning) that supplements trend analysis and can be used as a foundation for defining wild cards.

Web 3.0

Web 3.0 refers to the next generation of the World Wide Web that is characterized by advanced technologies and a more intelligent, interconnected, and decentralized internet. While there is no universally agreed-upon definition, Web 3.0 is often envisioned as a set of emerging technologies and paradigms that aim to enhance the way information is accessed, shared, and processed on the internet.

Index

About the Author

Phil Balagtas is a design leader with over two decades of experience creating innovative digital products and services. Working for companies such as General Electric and McKinsey & Company, he has served a variety of sectors, such as aviation, healthcare, utilities, consumer electronics, and retail, to name a few. As a consultant, he has helped transform businesses through cultural and digital transformations, trained teams on Design Thinking, and designed and shipped digital products for B2B and B2C organizations.

He is also the founder of the Design Futures Initiative, a nonprofit based in San Francisco focused on the advancement of Futures Thinking—principles and methodologies for assessing the threats and opportunities of the future. Through this initiative, he has worked with the United Nations and the Boys & Girls Club of San Francisco and has also founded a global network of Futures Thinking chapters called Speculative Futures. Working as a futures expert, evangelist, and consultant, he has been studying and applying futures to a variety of business and organizational contexts in an attempt to integrate it into the design practice and make the approach more practical, accessible, and viable for business and product strategy.

Colophon

The cover illustration is by Susan Thompson. The cover fonts are Gilroy and Guardian Sans. The text fonts are Minion Pro and Scala Pro; the heading and sidebar font is Benton Sans.